AN INTRODUCTION TO ENTOMOLOGY

About the Authors

Dr. P. D. Srivastava born at Azamgarh (Uttar Pradesh), had his schooling in his home town after which he moved to Allahabad for higher education. After doing his B.Sc., M.Sc. and D.Phil from the University of Allahabad, he joined the Indian Agricultural Research Institute, New Delhi (IARI). At IARI, for a few years, he worked on Insect Physiology and Locust Toxicology. However, he had to change his line of work in 1965 and started working on land snail pests especially the giant African snail (the biggest-sized land snail in the world) both at IARI, New Delhi and on the Andamans. He continued to work on this pest till his retirement in December, 1980.

The author was selected as a member of Post-Graduate Faculty, from the very inception of the Post-graduate School at IARI in 1958. He took M.Sc. and Ph.D. classes and guided a number of students leading to Master's and Doctorate degrees.

He has published 51 (fifty one) research papers in Indian and Foreign journals. So far he has authored three books, *Economic Zoology 1977; Faslon Ke Shatru—Bharat Ke Thal Ghonghe* (in Hindi) published by ICAR, New Delhi 1992 and *Problem of Land Snail Pests in Agriculture—A Study of the Giant African Snail* published by Concept Publishing Co., New Delhi, 1992.

He is one of the editors of *Agricultural Entomology,* Vols. I and II. He has also edited another book *Crops and Weather* as an External Editor and published by ICAR, New Delhi in 1992.

The author is a Fellow of the Entomological Society of India, New Delhi, Life Member of (i) National Academy of Sciences, Allahabad and (ii) All India Scientific Writers' Society, New Delhi. He is also recipient of one of the 1984 Statesman Awards for Rural Reporting.

Dr. R.P. Singh born and educated upto B.Sc. at Balrampur (Uttar Pradesh), obtained his M.Sc. degree from Gorakhpur University, Gorakhpur. After doing M.Sc. he joined the Indian Agricultural Research Institute (IARI) in 1964 from where he obtained his Ph.D. degree. He worked initially on Biological Control of Insect Pests and later shifted to Insect Toxicology. At present he is working on Botanical Insecticides for management of crop pests.

He was selected as a member of Faculty of the Post-graduate School of IARI in 1979 and since then has been taking Post-graduate classes and guiding students leading to Master's and Doctorate degrees.

He has published 54 research papers in Indian and foreign journals. The author is Associate Editor of *Neem Newsletter* published quarterly by IARI. He is co-editor of an IARI publication *Neem in Agriculture.* He is also a member of the Advisory Board of Pakistan Journal of Entomology. He is also one of the editors of *Proceedings of the World Neem Conference.* He has attended several International Conferences both in India and abroad. He is a fellow of the Entomological Society of India, New Delhi, life member of the Society of Pesticide Science, New Delhi and Allelopathy Society of India, Hissar (Haryana).

AN INTRODUCTION TO ENTOMOLOGY

P.D. SRIVASTAVA
Former Senior Scientist (Entomologist)
Indian Agricultural Research Institute
New Delhi

and

R.P. SINGH
Senior Scientist (Entomologist)
Indian Agricultural Research Institute
New Delhi

Foreword by
M.D. PATHAK
Director General
U.P. Council of Agricultural Research, Lucknow

CONCEPT PUBLISHING COMPANY, NEW DELHI-110059

ISBN 81-7022-624-4 (PB)

First Published 1997

Published and Printed by
Ashok Kumar Mittal
CONCEPT PUBLISHING COMPANY
A/15-16, Commercial Block, Mohan Garden,
NEW DELHI-110059 (India)
Phones : 5504042, 5554042
Fax : 091-11-5598898

FOREWORD

It has been my pleasure to know Dr. P.D. Srivastava since 1958. He has been an outstanding teacher and researcher in Entomology and has played a key role in the development of Agricultural Research and Education in India. He has been a highly respected and accomplished teacher and many of his students are occupying key positions nationally. I am very pleased to see the Book *An Introduction to Entomology* prepared by Dr. P.D. Srivastava and Dr. R.P. Singh and in which they have brought in their many years of teaching experience in various aspects of Agricultural Entomology. The book is nicely illustrated. It has a very extensive coverage of Entomology and I hope that students, teachers, and researchers will all find it a very valuable publication.

August 30, 1995

M.D. PATHAK
Director General
U.P. Council of Agricultural Research
Lucknow

PREFACE

When the Indian Agricultural Research Institute (IARI), New Delhi was recognised by the University Grants Commission as a 'Deemed University' in 1958 and Post-Graduate classes were started at IARI, Late Dr. S. Pradhan, the then Professor of Entomology introduced a course 'Introductory Course in Entomology'. This course included some non-traditional topics usually not given in books.

The present publication introduces the subject of Entomology through Arthropoda, numerically the biggest phylum of the animal kingdom. It deals with the awareness of the science of Entomology amongst ancient Indians as well as the development of this subject in modern India. It also deals with the factors responsible for the dominance of insects in the animal world.

Harmful aspects of insects have also been given due attention. Insects affecting crops, cattle, poultry and man have been listed and the nature of damage in each case has been described. How intensification of agriculture has upset the balance of nature leading to increase in pest population has also been discussed. The book also deals with the latest methods of insect pest management including the use of plant products as pesticides, the use of biological control agents and integrated pest management.

Role of beneficial insects in human welfare has also been dealt with adequately. Certain crops would give no yield, unless those are cross pollinated by insects. There are a number of crops whose yield increases many times because of cross pollination by insects. Thus insects are helpful in increasing agricultural production also. Honey, silk and lac are insect products. All these along with other beneficial aspects of insects have also been dealt with.

It is hoped that the book will find favour with the teachers and students alike. Any suggestion from any quarter would be given due consideration and may be incorporated whenever the reprinted/revised edition comes out.

The authors are grateful to the Director, Indian Agricultural Research Institute (IARI), New Delhi for kindly according permission to write the book and to the Head, Division of Entomology, IARI for kindly providing facilities including consultation of literature in the Divisional Library. The authors are also thankful to Head, Library Services, IARI for permitting to consult literature in the Library.

The authors also wish to express their thanks to Dr. (Mrs.) S. Singh for her help in the preparation of the manuscript dealing with the stored grain pests, Dr. K.L. Doharey for his help in the preparation of manuscript and Sri Pradeep Kumar Kataria for his help at different stages of the preparation of manuscript of the book.

P.D. SRIVASTAVA
R.P. SINGH

CONTENTS

LIST OF FIGURES

LIST OF TABLES

1

INTRODUCTION

A. POSITION OF ENTOMOLOGY IN SCIENCE

The word 'Entomology' is of Greek origin, Gr. *Entoma* meaning insect and Gr. *Logia* meaning discourse. Thus the science of Entomology is a discourse on insects. There can be several aspects of such a discourse. Thus, a science which gives an account of insects is Entomology and Entomology which deals with the economic role of insects is known as Economic Entomology i.e., the study of insect pests and their management as well as the useful insects and their role in human welfare.

The word 'science' has been derived from the Latin word *Scientia* meaning 'to know'. It emerges from the progressive activities of man to the extent that new concepts arise from experiments and observations and these new concepts lead to future experiments and observations. Different sciences are like branches diverging from different points from the trunk of a tree. The branch of biological science dealing with animal life is known as Zoology, of which Entomology is one of the most important branches of study.

B. NEED FOR THE STUDY OF ENTOMOLOGY

Insects are of immense importance to man because of their role (a) as pests in the field (e.g. borers, leaf eaters, sap suckers etc.) and in stores (e.g., *Trogoderma granarium, Sitophilus oryzae* etc.); (b) as predators and parasites of crop pests (e.g., *Coccinella, Cicindella, Syrphus* grubs, *Trichogramma* spp., *Microbracon* spp., etc.); and (c) as pollinators of flowers of many crops where fruit setting may not take place without the help of insects (e.g., closed flowers of *Ficus* sp. unisexual flowers of cucurbits, papaya etc., leguminous flowers of subfamily Papilionaceae). Bees had to be imported in Australia for cross fertilization of red clover flowers.

Insects cause damage to household materials like furniture, clothes, books etc. (e.g., crickets, cockroaches, silverfishes, dermestid grubs etc.)

Insects are also of medical and veterinary importance because of their role as vectors of certain diseases in man, cattle horses, fowls etc. (e.g., *Anopheles* spp. the mosquitoes; *Musca* spp., the houseflies, *Pediculus* spp., the body lice, *Phlebotomus* spp., the sandflies, *Glossina* spp., the tse-tse flies, *Boophilus annulatus,* the cattle tick etc.)

Certain insects like dung rollers act as scavengers. For this purpose Indian dung rollers also were needed in Australia to clear the dung of placental mammals (Menon, 1972).

Insects are also of industrial importance e.g., silkworm, lac insect, honey bee, cantharid beetles etc.

C. DIFFERENT ASPECTS OF ENTOMOLOGICAL STUDIES: FUNDAMENTAL, ACADEMIC, APPLIED AND DEVELOPMENTAL

1. *Fundamental/Basic* aspect of study deals with the discovery of new principles or facts. To work out the life history of an insect is a fundamental study. It may be either of academic (if it is of an insect of no economic importance) or of applied interest (if it is of a pest or useful insect).

2. *Academic* aspect of study is the pursuit of knowledge for the sake of knowledge to satisfy the curiosity of the worker. The knowledge thus gained may or may not be of any utility to humanity. For an Academician, rare insects are very valuable to satisfy his curiosity whereas a common insect found in abundance may not be of any special interest to him.

3. *Applied* aspects are those aspects of study which are pursued with a definite utilitarian aim in view, immediately/in near future/in distant future.

There is no difference in the 'technique' of academic and applied studies. The difference lies in the aim of study. Applied work is carried out with the ultimate aim of utility of benefit to humanity. While a rare insect is of very great value to an Academician, it is of no significance to an Applied Entomologist. If it happens to be a pest, it may not do any appreciable harm hence it would be very uneconomical to work out its management strategy; if it happens to be a useful insect, it may not be worth exploiting for the benefit of humanity because of its extremely low population.

After carrying out the fundamental study of the life history of an insect, to develop the technique of mass rearing for working out an effective and economical control measure or to exploit the mass-reared insect population (e.g., of silkworm) to get the maximum benefit out of it would be then of applied value.

4. *Fundamentals of academic* studies are discoveries of the new principles, facts, phenomena or theories which are of purely academic interest not related with human welfare.

5. *Fundamentals of applied* studies are the discoveries of those principles, phenomena, facts or theories which may be of use to man in one way or the other and may finally lead to investigations of applied value. Study of the mode of action of an insecticide effective against any particular pest is actually the fundamental of the applied aspect of that particular insecticide.

Anatomical, morphological, taxonomic and physiological studies are all fundamental studies. If these studies are carried out in relation to insects of no applied value these studies would be covered under fundamentals of academic studies whereas these very studies if carried out in relation to a pest come in the purview of fundamentals of applied studies.

6. *Developmental (operational)* studies are those in which applied researches are exploited for the benefit of humanity. Mode of their exploitation—implementation for the benefit of humanity is the developmental aspect of study e.g., treatment of sorghum seeds with carbofuran protects the crop from shootfly damage (Jotwani, *et al.* 1971, Sukhani, 1986); treatment of wheat seeds with aldrin protects the crop from termites and aldrin treatment of gram seeds protects the crop from termites as well as cutworms. These investigations are of great applied value. How to treat the seeds adequately so that it may be most effective and economical is the developmental aspect of these applied findings. Likewise a particular pesticide may be effective against a particular pest, but to work out as to which formulation is most effective/economical in actual practice is a developmental study.

Thus, we find that practically all these studies are interdependent, interwoven and interconnected and go hand in hand.

D. NEED FOR CORRECT IDENTIFICATION

The proper identification of pest and its natural enemies is essential since improper identification could send a biological control scientist on a foreign expedition to Europe instead of South America to look up for a natural enemy. A survey for the native home of sugarbeet leaf hopper *Eutettix tenellus* was carried out in South America for its natural enemies. Later, it was found that this species actually belonged to the genus *Circulifer,* an entirely different group of leafhoppers and that its native home was in arid region around Mediterranean. Biological control workers then searched in that area and succeeded in finding several natural enemies (Bartlett, 1964). Similarly when mass-reared parasites are to be introduced in the field for the control of a certain pest, it has to be checked up whether the field population of pest in the field where parasites are to be released is large or not, as in the absence of suitable host the entire introduced parasite population is likely to die.

E. ECONOMIC RELATIONSHIP OF INSECTS WITH PLANTS AND NON-INSECT PESTS

Taxonomically class Insecta is far away from other forms of life (except Arthropoda). However, economic relationship between insects and plants on the one hand and insect pests and non-insect pests on the other is very close.

Since most of the insects feed on plants, their damage to plant life is enormous. In this way economic relationship between insects and plants is very close. In the same way since many of the non-insect pests like mites, snails, slugs, birds and rodents etc., cause severe damage to crops, economically they are so closely related that they cannot be allowed to keep on damaging the crops simply because they are not insects; their number also has to be brought down by pest management below the economic injury level. This is why practically all the Entomology sections/departments at the agricultural centres/Agricultural Universities are carrying on investigations on mites, nematodes and rodents also. At the Indian Agricultural Research Institute, New Delhi, researches on mites, rodents and nematodes are also going on. At this Institute research on land snails was started in 1964 and went on upto 1980.

In recent years Agricultural Universities have started creating separate Departments of Zoology to devote more attention to non-insect pests. Indian Council of Agricultural Research is also financing a National Programme on Rodent Pest Management, being coordinated by the Central Arid Zone Research Institute, Jodhpur.

Thus an Economic Entomologist, in practice, has to be an Economic Zoologist also in order to be able to protect the crops and their produce from both insect as well as non-insect pests.

F. RELATIONSHIP OF INSECTS WITH OTHER ARTHROPODS

Numerically phylum Arthropoda is the largest phylum of the animal kingdom. It is characterised (1) by being bilaterally symmetrical (i.e. if cut into two parts lengthwise, both the halves correspond in structure). (2) mouth and anus situated at the anterior and posterior body ends respectively, (3) presence of ventral nerve cord and dorsal brain, (4) presence of haemocoele instead of coelom, (5) body externally segmented and some or all presence of bearing paired, jointed appendages. Segmentation of the body is shared by annelids and some helminths e.g., earthworm, leeches and cestodes. Presence of paired jointed appendages is shared by many other groups of animals including man and (6) presence of chitinous exoskeleton. The other groups possessing exoskeleton are Mollusca and Echinodermata. The molluscan exoskeleton (shell) and the exoskeleton of echinoderms are calcareous.

G. ANCESTRY OF INSECTS

There are a number of theories of the origin of insects; one of these is the crustacean theory. Although crustaceans are aquatic arthropods and insects are mostly terrestrial/aerial, similarities do exist between certain body parts. Head and thorax of insects correspond to the crustacean cephlothorax. Appendages of cephalothorax of crustacea and of head and thorax of insects are homologous (Pruthi, 1969) as given below.

1. A pair of antennules of Crustacea are homologus to a pair of antennae of Insecta.
2. A pair of antennae of Crustacea are not represented in Insecta.
3. Mandibles of Crustacea are homologous to the mandibles of Insecta.
4. First pair of maxillae of Crustacea is homologous to the first pair of maxillae of Insecta.
5. Second pair of maxillae of Crustacea is homologous to the second pair of maxillae/labium of Insecta.
6. First, second and third pair of maxillipeds of Crustacea are homologous to the first, second and third pair of thoracic legs respectively of Insecta.

Myriapoda, like Insecta possess a single pair of antennae which are homologous to the antennule of crustacea.

As stated earlier, insects are mostly aerial and terrestrial. They resemble Myriapoda in the possession of mandibles and tracheae although the tracheal system of both the classes appear to have developed independently. Insects can be separated from other terrestrial arthropods by the division ofbody into three distinct regions, presence of a single pair of antennae, three pairs of thoracic legs and one or two pairs of wings.

The wingless condition of Apteygota is considered to be of "primitive and not of secondary origin". Although amongst the Apterygota only Thysanura "approach most closely to the Perygote insects" and are likely to be near the main line of pterygote evolution, "the Diplura are of considerable phylogenetic interest because of their resemblance to Symphyla (Imms, 1936).

Thus Thysanura and Diplura represent the most primitive insects when these are compared with the myriapod classes. Symphyla are found to exhibit most of the "essential structural features required of an ancestral form". Thus the presence of a Y-shaped epicranial suture, two pairs of maxillae, styli and eversible sacs on the abdominal segments, presence of anal cerci and Malpighian tubules are common to both Symphyla and *Campodea* (Diplura). Similarly *Anajapyx* (Diplura) and Symphyla both possess abdominal styli and eversible vesicles and their anal glands open on the cerci. Symphyla and *Anajapyx* also agree in the structure of their mouth parts, antennal musculature and legs (Imms, 1965). The main objection against the symphylan origin of insects is the "position of gonopore" which in insects is near the posterior end of abdomen but in Symphyla, it is situated anteriorly. However, Tiegs (1940, 1945) is of the view that progoneate condition of Symphyla "is a secondary specialization and probably the insectan stem arose from Symphyla before this secondary specialization took place.

The Trilobitean, Crustacean and Collembolan theories of insect origin are out of date and unacceptable to the Zoologists/Entomologists. The Symphylan theory of insect origin appears to be generally acceptable (Fig. 1).

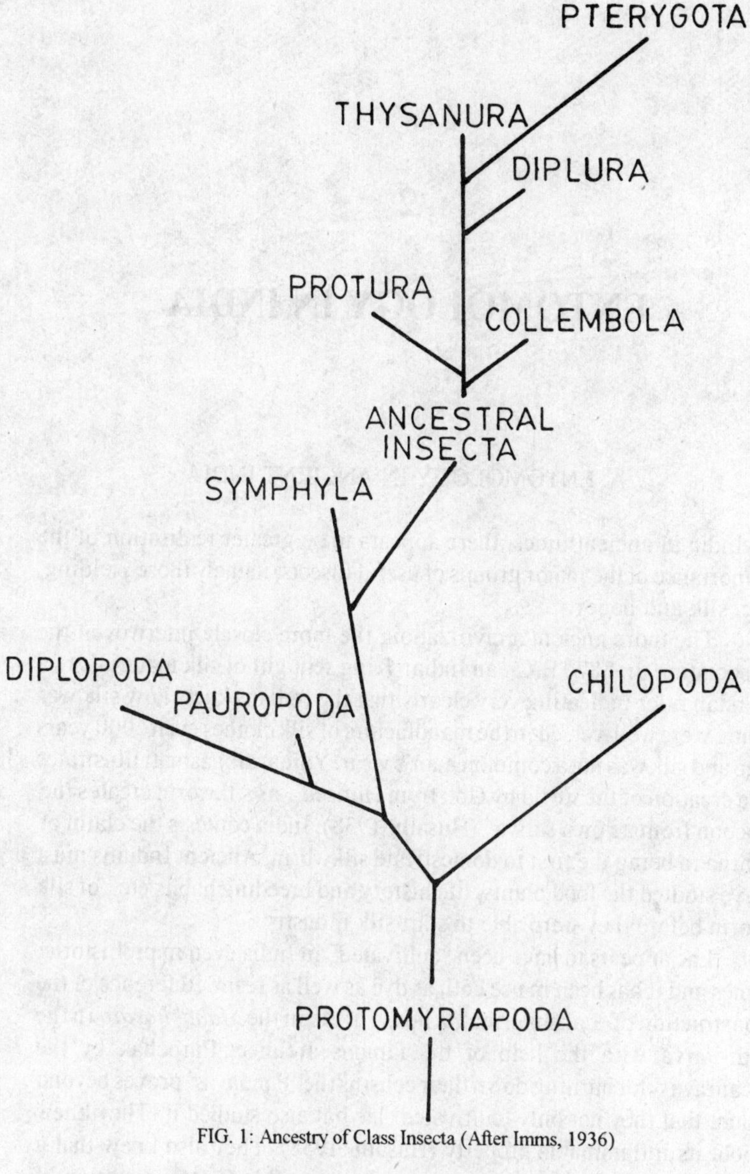

FIG. 1: Ancestry of Class Insecta (After Imms, 1936)

2

ENTOMOLOGY IN INDIA

A. ENTOMOLOGY IN ANCIENT INDIA

In India in ancient times, there appears to be greater realisation of the importance of the major groups of useful insects, namely those yielding, lac, silk and honey.

The more ancient a civilization, the more closely interwoven the association. In 3870 B.C., an Indian King sent gift of silk materials to a Persian ruler indicating very clearly that the Indians knew how silk was spun, were well-versed in the manufacture of silk clothes even 6000 years ago and silk was not a common man's wear. Yajnavalkgasmati illustrates the creation of the world by God from Himself "as silkworm creates the cocoon from its own saliva" (Husain, 1938). India contests the claim of China in being the first to domesticate silkworm. Ancient Indians must have studied the food plants, life history and breeding habits etc., of silk worm before they were able to start silk industry.

Lac appears to have been "cultivated" in India even in prehistoric times and it has been in use both as dye as well as resin. Reference of the construction of 'Lakshagriha', a house of lac in the *Mahabharata* in the Adi-Parva with the help of the famous architect Purochak by the 'Kauravas' for burning down their cousins the 'Pandavas' proves beyond doubt that they not only 'cultivated' lac but also studied it. They knew about its inflammable property (Husain, 1938). They also knew that a large number of insects together were responsible for the secretion of 'laksha' (lac). *Laksha* in Sanskrit and 'lakh' in Hindi means hundred thousand (Pradhan and Jotwani, 1964).

Honey was the only substance in general use for sweetening food until 17th century A.D. There are copious references to *Madhumakshika* (honey bee) and honey in ancient Sanskrit literature, "honey being used as food and in rituals" (Husain, 1938). *Rigveda,* one of oldest available records dating back to 3000-5000 B.C. mentions honey as the choicest food of gods, and refers to bees as harbingers of wealth, prosperity and sweetness in life. It continues to form an essential ingredient in various rituals and religious ceremonies. Even now honey forms an integral part of the diet of many Indian Tribals e.g., Oongies of Andaman and Nicobar Islands living in jungles far away from the modern civilization.

There is also ample evidence to show that "ancient Indians devoted considerable attention to the discovery of medicinal properties of insects." This could not have been possible without" careful researches and extensive and elaborate observations continued over hundreds of years." In ancient books of medicine, there are abundant references to "insect tormentors of man" and various remedial and preventive measures have been suggested (Husain, 1938).

The use of 'Chauri' must have originated 'as a measure of warding off the attention of flies'. Burning incenses in temples and during religious ceremonies probably had its origin, as repellents against mosquitoes and other insects seeking refuge in the dark, cool and humid interiors of the temples. There is frequent mention in Sanskrit medical works of the repellent properties of incenses and there is reference in *Matsyapurana* of an incense "so potent that where it is burnt no insect will be seen there, neither frogs nor snakes" (Husain, 1938).

There are hymns in *Atharvaveda* concerning control of insect pests of crops. These hymns also clearly indicate that "although the pest problems were really serious the knowledge about them and their control was very inadequate" (Pradhan and Jotwani, 1964). These hymns, however, give an insight into the variety of pests that had to be destroyed and the thoroughness of destruction that was aimed at (Husain, 1938). Reference to *patanga, pipilika, makshika* etc. in ancient Sanskrit literature shows that ancient Indians knew about insects. *Amarkosha* refers to flies, beetles, bees, hornets, moths, ants and glow-worms.

According to a story in *Mahabharata*, sage Mandavya who used to collect and pin insects as a hobby had to pay the penalty for it by being nailed on a crowbar (Ayyar, 1939).

In ancient Indian literature the 'black bee' was described as 'a stinging insect with six legs' and was given the name 'shat-pada' meaning six legged (Amarkosha, Ist century A.D.), while P.A. Latreille coined the term Hexa-poda for insects as late as in 1825 A.D. (Husain, 1938).

Social insects like the honey bees appear to have been studied thoroughly. What man learnt later, social insects practised earlier.

(1) Manu, while dividing the Hindu society into castes based on division of labour appears to have the knowledge of the existence of caste system amongst honey bees where each caste has a particular work to do.

(2) The idea of poisoned arrows by man for hunting appears to have emerged from the stinging Hymenoptera (including honey bee).

(3) While the idea of socialism emerged from the study of social insects, the idea of monarchy also, the monarch having better amenities of food and dwelling, appears to have been borrowed from social insects.

(4) The idea of slavery, keeping slaves only to work may have its origin in the sterile female workers of honey bees which take care of the honey bee queen, not allowing her to do any work except to reproduce.

(5) Reproduction in honey bee is "planned reproduction not surplus, just enough for the colony and if any are surplus they are killed. We are talking about" planned reproduction in man now in the 20th century.

(6) More than one queen is not permitted to exist in one honey bee colony. Whenever there is more than one queen either it is killed or the old queen followed by a number of loyal workers leaves the hive, making room for the younger queen to take over the colony. The old queen establishes a new home elsewhere. This ensures undivided loyalty to the queen and the colony.

(7) When a male has mated with the queen bee, it is killed so that it may not mate with another female. Not only this, it also becomes incapable of mating with another female as the aedeagus of the mating male, after fertilizing the queen, gets broken and detached from the body. This is a form of monogamy.

(8) Nuptial flight of the queen who is followed by a large number of males appears to resemble the institution of *Swamvara* where a large number of men tried to get the hand of a woman but only one was successful, the rest had to go back.

(9) The perfect hexagonal cells of bee comb and its vertical parallel walls are so accurate that this precision is not

comparable to any structure prepared by man. Thus science of geometry, the oldest science of mathematics has its origin in the study of the precise and perfect cells of social insects.

All this further indicates that the ancient Indians did have knowledge of Entomology more so of the useful insects like lac insect, silk worm and honey bee.

The entomological studies in Europe can be traced back to the time of Aristotle (384—320 B.C.) long after the Indian learning had reached a very advanced standard (Husain, 1938). However, we are not aware of the entomological observations of the Indian counterparts of Aristotle (Husain, 1938). Observations about the behaviour of ants have been salvaged by Moses (1928) (quoted by Husain, 1938) from folk lore. Such researches in the field, yet completely unexplored, may yield very useful results.

Development of Science of Entomology in India

The foundation of Indian Entomology was laid by amateur European Naturalists.

Earliest Record of Indian Insects

The earliest available record of Indain insects is that of Linnaeus who included 12 Indian insects in the 10th edition of his famous book *Systema Naturae* published in 1758 A.D. Thus the foundation of Taxonomic Entomology in India was laid by Linnaeus more than two centuries back. Next record is of Fabricius who received a large number of insect collections from some Christian missionaries and officials of East India Company and included more than 1000 Indian insect species in his publication *Entomologia Systematica* between 1792 and 1798 (Pradhan and Jotwani, 1964).

Work on Indian insects in India was initiated by S.G.Koenig, a pupil of Linnaeus. He first came to India as a Medical Officer to the Tranquebar Mission but later was appointed as an Official Naturalist to the Nawab of Arcot for some time. He collected insects of Caromandal Coast and sent to Linnaeus and Fabricius for identification (Pradhan and Jotwani, 1964). He published his own studies on 'Termites of Tanjore district' in 1779, the translation of which in English was published much later in 1921 in the Proceedings of the Fourth Entomological Meeting, Pusa. Other Indian Entomological publications of 18th century are of

Kerr on lac insect published in 1782, 1790 and 1800. Valuable observations on silk worms and their culture in some parts of South India and a monograph on cochineal insects were also reported during this period. Based on the insect collections of Francillon, Drury, Macleay, Banks and Donovan's own collections, the first general work on Indian insects started publication in 1800 as *Natural History of Insects in India* (Husain, 1938). Thus far, the science of Entomology had progressed up to the end of 18th century. This period can be said to be a period of Natural History Studies of Indian Insects.

B. DEVELOPMENT OF ENTOMOLOGY IN INDIA IN THE 19TH CENTURY

In the 19th century Entomology made tremendous progress. Donovan's *Natural History of Insects in India* was revised by Westwood and published in 1842. Westwood's own classical work appeared in 1847 as *Cabinet of Oriental Entomology* (Husain, 1938).

A paper on the "Entomology of the Himalayas and India" by F.W. Hope appeared in 1840 in the *Madras Journal of Literature and Science* and a paper on "Indian Ants" was published in 1893 (Pradhan and Jotwani, 1964).

The first scientific society to be established in India, was the "Asiatic Society of Bengal" at Calcutta in 1785. This society encouraged insect collection and exhibition of Indian insects and published entomological literature. About 1836 Lord Aukland, Patron of the society arranged informal social gatherings to encourage scientific knowledge and its dissemination by arranging and exhibiting insect collections. Throughout the 19th century taxonomic study of insects received a good deal of attention. The entomological publications contained lot of information based on the collections mostly of amateur foreigners and identified by European specialists. In 1875, ninety years after the establishment of Asiatic Society, the Indian Museum was established and the entire collection of the Asiatic Society of Bengal was handed over to the Museum (Husain, 1938). In 1884, when Cotes (quoted by Husain, 1938) was appointed as the first Entomologist to the Museum, insect collection at the Museum was only 20,000 but within the next ten years (1884—1894) the collection swelled to four times. It was sorted out and classified by Cotes.

Cotes published the results of his as well as of other workers' investigations on applied entomology in the first three volumes of the *Indian Museum Notes*. The publication of the *Indian Museum Notes* was started in 1889 and financed by the Government of India (Pruthi, 1969). This can be called as the birth of applied entomology in India. Since Cotes left India in 1896, the publication of fourth and fifth volumes of the *Indian Museum Notes* was arranged by Barlow and Peal (Husain, 1938).

C. DEVELOPMENT OF ENTOMOLOGY IN INDIA IN THE 20TH CENTURY

Lionel de Niceville was appointed as the first Entomologist to Government of India in 1901 with headquarters at the Indian Museum. He was required to take observations on crop pests in the field in actual areas of their occurrence. He had gone to study the Mosquito Blight of Tea at Darjeeling where he got fever and died in December 1901. After his death E.P. Stebbing officiated as Superintendent of the Indian Museum for some time. During 1903 he published Part I of the sixth volume of the *Indian Museum Notes*—the last publication on Economic Entomology from the Indian Museum. H. Maxwell Lefroy was appointed as Entomologist to Government of India in 1903, in place of Lionel de Niceville but was posted at Surat to study the cotton pests.

Lord Curzon's Viceroyalty (1898—1905) marks the beginning of a new era of growth of Agricultural Departments both at the centre and in the provinces. The famine of 1899-1900 convinced him that the Government of India must pay urgent attention to agriculture (Randhawa, 1979).

In 1905, when Bihar was a part of Bengal, the Government of Bengal proposed to utilize a large government estate at Pusa in Bihar as a 'Provincial Research Station and College'. The Government of India with the full concurrence of the Bengal Government, took over the estate for the purpose of starting an Agricultural Research Institute and Experimental Farm and Agricultural College (Randhawa, 1979).

On April 1, 1905 the Imperial (now Indian) Agricultural Research Institute was established at Pusa (Bihar) with a section of Entomology in addition to some other sections. Agricultural Colleges were reorganised or started at Pune, Kanpur, Nagpur, Lyallpur and Coimbatore. Lefroy was then transferred from Surat to Imperial Agricultural Research Institute, as the First Imperial Entomologist. In 1907, a post of Second Imperial Entomologist was created but later it was redesignated as Imperial Pathological Entomologist. Lefroy was the pioneer of Indian Entomology. He published Indian Insect Pests in 1906 and in 1909 Indian Insect Life which are indeed testimony to the knowledge, energy and industry of Lefroy and his co-workers (Husain, 1938). In 1915-16 he wrote a comprehensive report on sericulture. In 1913, T.B. Fletcher succeeded him. During his tenure, Insect Taxonomy received greatest attention. He build up a valuable insect collection especially of Lepidoptera and build up a splendid library. So far there was no proper forum for the Entomologists to meet and exchange their views. Fletcher organised five All India Entomological Meetings at Pusa between 1915 to 1923 and published their proceedings.

The second scientific society established in India was the Bombay Natural History Society in 1883—a century after the establishment of the Asiatic Society. In 1888, this society started its own publication as the *Journal of Bombay Natural History Society.*

On account of the availability of these media *(Journal of Asiatic Society of Bengal, Indian Museum Notes* and *Journal of Bombay Natural History Society)* for entomological publications, in addition to the recording of the occurrence of insect pests, workers also started working on the biology, bionomics and host records. In July 1916, by separating the Natural History section of the Indian Museum, the "Zoological Survey of India" was established. This institution has since then been working on practically all groups of animals including insects and is one of the three major centres having Indian Insect Collections.

In 1909, F.H. Gravely was appointed as Assistant Superintendent in Entomology at the Indian Museum but in 1920 was transferred to Madras and the post remained vacant for about three years. In 1923, B.N. Chopra took over followed by H.S. Pruthi in 1925. Dr. Pruthi remained Incharge of the section of Entomology of the Zoological Survey of India upto 1934, when he was transferred to Imperial Agricultural Research Institute, Pusa to take over from Fletcher.

To consolidate the available information on Indian insects scattered in different publications, the Government of India commenced publication of the *Fauna of British India* series from London in 1892, the first of the series being on the 'Moth Fauna of India.' Likewise the first 'Catalogue of Indian Insects' on Testtigidae *(Acridiidae)* was published in 1921 followed by the publication of catalogues of other insects.

Tea Association was formed in 1905 at Kannykorrien (Assam) where an Entomological Laboratory was established under the charge of C.B. Autram. In 1911, Autram resigned and was succeeded by E.A. Andrewes. Later in 1913, the headquarters were shifted to Tocklai.

Simultaneously, with Agricultural Entomology, insects attacking forest trees were also being studied and there are numerous references to forest insects in the Insect Museum Notes. In 1898, E.P. Stebbing, published a booklet of forest insects and in 1900 he was appointed as

Imperial Forest Entomologist. The Imperial Forest Research Institute was established at Dehradun in 1906 and Stebbing was appointed there as Forest Zoologist. He stayed in this post for over two years and left India in 1909. In 1911 A.D. Imms was appointed in place of Stebbing. Within the short period of just 16 months, Imms remained at the Forest Research Institute, he "laid the foundation of an authentically identified insect collection, a good library and an adequate laboratory." (Husain, 1938). In 1913, C.F.L. Benson was appointed as Forest Zoologist; later the designation was changed to Forest Entomologist. In 1914, Stebbing's excellent book on Indian Forest Insects appeared. During Benson's term, Forest Entomology made good progress and a Systematist was added to the staff. M. Cammeron was appointed Systematic Entomologist in 1920. He was succeeded by J.C.M. Gardner in 1923.

In the first quarter of the present century many important developments took place. The establishment of the Indian Agricultural Research Institute had a definite impact on the states/provinces which started organising their own departments of agriculture. The entomological work related to agriculture is reported to have started in 1902—1903 in Madras, in 1905 in Punjab and in 1906 in U.P. In 1909, there were only 13 Entomological Assistants in the provinces and 16 in 1910,—five for teaching and 11 for field work (Husain, 1938). However, regular posts of Government Entomologists were created first by the then princely state of Mysore which appointed Coleman as Entomologist and Mycologist in 1908. Madras appointed T.B. Fletcher as Government Entomologist in 1912, Punjab appointed M. Afzal Husain in 1919 and U.P. appointed P.B. Richards in 1921 followed by other states. Punjab appointed an Assistant Professor of Entomology in 1908, much before the appointment of a state Entomologist. Now all the states have their own departments of agriculture with full-fledged sections/departments of Entomology headed by academically qualified Entomologists (Pruthi, 1969).

The Indian Central Cotton Committee was established in 1921 and it started financing projects on cotton pests at Bombay (Surat), U.P. and Punjab. In the same year a Lac Commission was appointed on whose recommendation Indian Lac Research Institute was established in Bihar in 1925 although work on lac insect was already being done at the Imperial Agricultural Research Institute (Pusa-Bihar) by C.S. Misra and at Forest Research Institute, Dehradun by A.D. Imms and N.C. Chatterjee (Husain, 1938).

Insect Collections

Establishment of Imperial Agricultural Research Institute at Pusa, Indian Museum—Zoological Survey of India at Calcutta, Forest Research Institute at Dehradun with provision for researches in Entomology, All India Entomological meetings at Pusa, publication of its proceedings and a number of other entomological publications put Entomology in India on sound footing. Imperial Agricultural Research Institute, Forest Research Institute and Zoological Survey of India emerged as the three major centres housing insect collections of national importance, Imperial (now Indian) Agricultural Research Institute housing about 15,000 species of insects mainly of agricultural importance, Forest Research Institute housing about 20,500 species of insects mainly of forestry importance and Zoological Survey of India housing about 17,000 species mainly of general faunistic collections of insects, not excluding pests of agricultural crops and forests.

Medical and Veterinary Entomology

India has made valuable contribution to Medical and Veterinary Entomology. Ronald Ross declared in 1898 that certain "dapple-winged" mosquitoes were the carriers of malaria and by 1900 the Anopheline mosquitoes as vectors of human malaria had been conclusively implicated (Husain, 1938). About 1904 the King Institute of Preventive Medicine at Guindy (Tamil Nadu) started investigations on Kalaazar and the role of flea in the transmission of Plague. Thus in 1905 the Plague Commission was appointed. The Central Bureau of Malaria was established in 1910 and field laboratories were added to carry out research work on mosquitoes and Malaria. In 1927, it was renamed as Malaria Survey of India, in 1938 as Malaria Institute of India and in 1962 as National Institute of Communicable diseases (New Delhi).

The Indian Research Fund Association was organised to supervise and finance research enquiries into various human diseases and also on important entomological subjects closely connected therewith. Since then it has been financing the "entomological problems connected with the transmission and eradication of malaria, oriental sore, kalaazar, plague etc." (Husain, 1938). In 1913 Patton and Cragg published a *Text Book of Medical Entomology* and the Indian Research Fund Association started the *Indian Journal of Medical Research.* In 1933 Christopher's Volume of *Fauna of British India* on Anopheline mosquitoes dealing with 43 species and a year later Barraud's Volume on Megarhinini and

Culicini dealing with 272 species were published. In 1933 Sandfly fever became a problem for the military authorities in the then North West Frontier Province. In 1934 at Quetta, there was a severe outbreak of oriental sore. In 1907, Mackie discovered that the Asiatic Relapsing fever was transmitted by the body louse, *Pediculus.* At the School of Tropical Medicine, Calcutta in 1922 Strickland was appointed the first Professor of Medical Entomology. The need for the study of all insects of medical importance was stressed "without an eye for immediate results so that we discover all about them and then apply our knowledge for their control" (Husain, 1938).

Veterinary Entomology took birth in 1893, with the discovery of tick, *Boophilus annulatus* as the intermediate host of *Babesia bijemina* the causative organism of Texas fever of cattle, in the American region. In 1895 in the African region, the tse tse fly was discovered to be responsible for the transmission of *Trypanosoma brucei,* the protozoan parasite causing the Nagana disease of cattle. In India the discovery of the Surra disease of cattle, horses and camel being caused by the flagellate *Trypanosoma evansi* focussed the attention of Veterinary Officers on its possible insect vector/vectors. This was the beginning of Veterinary Entomology in India. Veterinary Entomology also received much help from those working on human diseases. Howlett (1919) pleaded for the creation of a separate organisation to deal with Medical and Veterinary Entomology in the country. At long last the post of Entomologist was created at the Imperial Veterinary Research Institute in 1933 and with the appointment of S. Sen to this post "the first brick" of the edifice of Veterinary Entomology in India was laid.

Thus more or less simultaneously towards the end of 19th century and in the beginning of the present century economic importance of insects in the fields of agriculture, forest, human and animal health began to be realised in India partly due to the developments in the country and partly due to developments in other countries.

Teaching of Entomology

In 1919-1920 Entomology was the 'Cindrella' of science and it had a minor place even in the agriculture degree course of this "predominately agricultural country". It had hardly any place in the curricula of Zoology in the Indian Universities. The position of Entomology in education was very unsatisfactory. The only training in the different disciplines of Agriculture including Entomology was available at the Imperial Agricultural Research Institute, Pusa in the form of training course started in 1908.

The third Entomological meeting held in February 1919 made the following recommendations. "Considering the great economic importance of insect pests to Indian Agriculture we recommend that all Agricultural Colleges should make provision for teaching of Entomology". (Husain, 1938). Since then the position has greatly improved. All the agricultural colleges which prepare students for the degree examinations in Agriculture give some training in Entomology also. The fifth Entomological meeting held in 1923 recommended to the Indian Universities that "the subject of Entomology be taught as of equal rank with other branches of Biological Science in the course of studies for the examinations of the universities for the degrees in Science including Agriculture." Imperial Agricultural Research Institute started a two-years Associate IARI Diploma course in 1923.

The fourth conference of the Indian Universities held in Bombay in 1939 also resolved that "the Universities should make adequate provision for the teaching of Entomology and give this subject proper place in the degree examination in view of the great educational value of Entomology and the immense economic importance of this science to Agriculture, Forestry, Public Health, Animal Husbandry and certain important industries in India". These developments enhanced the prestige of the science of Entomology in India. In accordance with this, courses in Entomology were introduced in a number of universities like those of Calcutta, Madras, Aligarh, Allahabad, Agra and Punjab. Calcutta University was the first to appoint a Lecturer in Entomology.

Till 1944 there were Master's degrees in Zoology with specialization in Entomology but there was no Master's degree in Agriculture in any subject. In 1944 for the first time Master's degree in Agriculture was started in several subjects including Entomology and the first batch of M.Sc. (Ag.) in Entomology came out in 1946. To this first batch from the Agriculture College, Kanpur belongs Dr. N.C. Pant, formerly Professor and Head of the Division of Entomology, Indian Agricultural Research Institute and later Director of the Commonwealth Institute of Entomology, London. He retired in 1984.

Alongside the introduction of M.Sc. courses in Entomology, Universities like Allahabad introduced courses in several agricultural subjects including Master's degree in Agricultural Zoology.

In 1948, there were only seventeen agricultural colleges in India and facilities for post-graduate research work were available only for 160 students (Randhawa, 1979).

The question of teaching Entomology in the Universities was

discussed at the Indian Science Congress Session of 1950 and 1951.

It was discussed again in the 1957 under the joint auspices of the section of Zoology and Entomology and the section of Agricultural sciences. It was pointed out that several universities had already introduced Entomology in their curricula but they should simultaneously and judiciously exclude proportionate number of existing topics.

In 1958 the Post-graduate School of the Indian Agricultural Research Institute was established and for the first time Institute was raised to the status of a Deemed University under section 3 of U.G.C. Act, with the right to award degrees in M.Sc. and Ph.D. The then Central College of Agriculture affiliated to Delhi University for the award of B.Sc. (Ag.) Hons. degree was closed down and the Associate IARI Diploma courses also were discontinued.

Radhakrishnan Commission on University Education (1948-49) recommended the establishment of 'Rural Universities' to supply skilled persons that would be needed by the country. This represented the germ of the idea that led to the establishment of Agricultural Universities in India (Randhawa, 1978). The two joint Indo-American Teams on Agricultural Education (1954-55 and 1959-60) recommended the setting up of Agricultural Universities in the states whereas the education commission (1964-66) headed by D.S. Kothari recommended the establishment of 'atleast one Agricultural University in each state'. It also recommended that all aspects of research on agriculture should be the concern of the agricultural universities. The first agricultural university to be established was Govind Ballabh Pant University of Agriculture and Technology at Pantnagar (Distt. Nainital) in 1960. Others came up later. At present there are twenty-seven agricultural Universities in India. Some states like Bihar, U.P., Maharashtra, Karnataka and Madhya Pradesh have more than one Agricultural University while a few have none so far.

In accordance with the recommendation of the National Commission on Agriculture, several research responsibilities of state department of agriculture have been transferred in several states to the agricultural universities.

Birth of ICAR

On the recommendation of the Royal Commission on Agriculture. Imperial (now Indian) Council of Agricultural Research was set up in 1929. The Council financed a number of Entomological Projects.

There was a Desert Locust cycle from 1926 to 1932 and in 1930,

ICAR sanctioned a scheme on the Survey of permanent breeding areas of the Desert locust and study of its biology. M. Afzal Husain was in charge of the scheme for 3 years and was succeeded by Y. Ramachandra Rao. This scheme continued for about 10 years; after its termination in 1939, the Government of India established a permanent Locust Warning Organization under the Imperial Entomologist of the Imperial (now Indian) Agricultural Research Institute, New Delhi, but with meagre staff at the outposts in the Deserts of Rajasthan, Sind and Baluchistan (Mekran) and with field headquarters at Karachi. The Locust Warning Organization was expected to keep a constant watch on the locust population and keep the Imperial Entomologist informed to enable him to warn the States/provinces of expected fresh outbreak of locust cycle and advise the states/provinces to undertake anti-locust measures. After about a year of the establishment of the Permanent Locust Warning Organization there was a fresh outbreak of locust cycle in 1940. Then it was realised that the meagre staff of the Locust Warning Organization was not enough to keep this pest in check. Thus in 1942 the Government of India sanctioned a Coordinated Anti-Locust Scheme with additional staff under which the Imperial Entomologist was required to control the pest in the desert breeding grounds in cooperation with the local authorities. As regards the control work in the cultivated areas, the states/provinces had to do themselves under the overall technical supervision of the Imperial Entomologist who was later designated as Director, Locust Control in India.

At the Imperial Agricultural Research Institute, Fletcher laid stress on Taxonomy and built up a very good insect collection library. Fletcher also started publication of the catalogue of Indian Insects. Fletcher was succeeded by Pruthi in 1934. After the shifting of the Imperial (now Indian) Agricultural Research Institute from Pusa to New Delhi in 1936 due to devastating earthquake of Bihar, the Institute started attracting more public attention and the Imperial Entomologist had to modify his research programmes with more leaning towards Economic Entomology.

At the Silver Jubilee Session of the Indian Sciences Congress (1938) a section of Entomology, independent of Zoology Section of the Congress was born but it was again merged with the Zoology Section in 1942 and a combined section of Zoology and Entomology was established. On 7.1.1938 the Entomological Society of India was born with Mian Afzal Husain as its first President. This Society started publication of its Journal, the *Indian Journal of Entomology* in 1939 with Dr. H.S. Pruthi as its first Chief Editor. The first article of the first issue of this Journal was on "Entomology in India—a retrospect" by T.V. Ramakrishna

Ayyar. This article will ever be a guide to the entomological workers (Pradhan and Jotwani, 1964).

The gradual academic recognition of Entomology had gone hand in hand with the development of fundamental aspects of this science. This is evident from the fact that when the post-war and post-independence expansion of Indian Agricultural Research Institute took place during late forties and fifties the Section of Entomology was raised to the status of Division and thus Head of the section of Entomology was redesignated as Head of the Division of Entomology and set up of the Division was thoroughly reorganised (Pradhan and Jotwani, 1964). A perusal of the review of Entomological Publications from 1938-1950 by Roonwal (1954 quoted by Pradhan and Jotwani 1964), shows that considerable attention was paid from theoretical as well as descriptive angle to embryonic and post-embryonic development, effect of temperature and humidity on the life of insects, population studies, evolutionary problems, basic studies in insect taxonomy, toxicology, parasitology and invention of a biometer. These were the significant achievements of the Indian Agricultural Research Institute from 1938-1950. Side by side developments in the researches leading to the control of pests, and researches on the useful role of insects especially the silkworm, lac insect and honey bee were also going on.

In 1968, Insecticide Act was passed and enforced from 1.8.1971 to regulate the import, manufacture, sale, transport, distribution and use of insecticides with a view to prevent the hazard to humans and animals.

Birth of Plant Protection Directorate

The Woodhead Famine Commission formed after the Bengal Famine of 1943, recommended in 1945, the establishment of a comprehensive organization to deal with pests and diseases of crops. Thus, in May, 1946, the Government of India established, on permanent basis, a Central Plant Protection Organization (now Directorate of Plant Protection, Quarantine and Storage) under the Ministry of Food and Agriculture, with headquarters at New Delhi, and Dr. H. S. Pruthi was appointed as Plant Protection Adviser to the Government of India. After the establishment of Plant Protection Directorate, entomology entered into a new phase of its development. The Plant Protection Adviser was also designated as Director Locust Control. The Locust Warning Organization hitherto under the Indian Agricultural Research Institute was transferred to the Plant Protection Directorate and formed a Division of the Directorate. The directorate also helped the states/provinces to establish their own Plant Protection Services.

The Plant Protection Scheme was started by U.P. and Punjab in 1947, by Madras in 1948 and by M.P. in 1949 followed by other states/ provinces. This came out to be economical and paying proposition. This was realised by the Entomologists and cultivators both and thus it became easier to convince the Government to expand the organization. Thus by 1957, the Plant Protection Directorate had set up Central Plant Protection Centres at Pathankot, Barabanki, Bilaspur, Amravati, Cuttack, Dharwad, Hyderabad, Gaya, Indore, Tiruchirapalli, Palampur, Ernakulum and Gauhati with headquarters at New Delhi (Pradhan and Jotwani, 1964). Gradually the State Governments established a network of their own Plant Protection Service centres in different parts of their states.

In 1966, a Central Plant Protection Institute was established at Hyderabad. This Institute is conducting a number of courses related to Plant Protection.

Pruthi retired in 1954 and was succeeded by K.B. Lal. After K.B. Lal other Plant Protection Advisers were Sardar Singh, S.N. Banerjee, K.D. Paharia and R.L. Razak. R.L. Razak went on deputation in July, 1993; since then V. Ragunathan is holding the additional charge of Plant Protection Adviser.

The Directorate was shifted to Faridabad in May 1968. At present the following 12 Divisions/Units are functioning in the Directorate (Anonymous, 1986).

1. *Locust Warning and Control unit:* It is spread over in the desert area and is divided into 4 circles, 9 zones and 34 outposts with field headquarters at Jodhpur in addition to a field station for investigation on Locust, functioning at Bikaner, since 1957.
2. *Plant Protection Division*: Presently 13 Plant Protection Stations are functioning in different states.
3. *Pest/Disease Surveillance Division:* It has 19 stations functioning in different parts of the country for forecasting the pest/disease situation mainly in the pest/disease-prone crops like paddy, sugarcane, potato and wheat.
4. *Biological Control Unit:* Eleven Central Biological Control Stations are functioning in different states. In addition, one parasite multiplication unit has been set up at Bangalore.
5. *Plant Quarantine Division:* After the establishment of the Plant Protection Directorate, the technical aspect of enforcement of Plant Quarantine measures of inspection/

fumigation was taken over from the customs and the first Plant Quarantine and Fumigation Station was established at Bombay in 1949. At present 29 Plant Quarantine and Fumigation stations—9 at sea ports, 10 at airports and 10 at land frontiers are functioning in the country (Anonymous, 1986).

6. *Implementation of the Insecticide Act*:* Implementation of this Act is being done by the Plant Protection Directorate. Under this act, there is a Central Insecticide Board, and a Registration Committee, (Table 2.1 to 2.4). A Central Insecticides Laboratory has been set up at Faridabad with branch laboratories at the Plant Quarantine and Fumigation Station, Bombay and the Central Plant Protection Training Institute, Hyderabad. Five Regional Pesticides Testing Laboratories also are being set up one each at Bombay, Calcutta, Chandigarh, Hyderabad and Kanpur with its headquarters at Faridabad (Anonymous, 1986).
7. *Storage Unit:* A scheme is in operation to collect information on the pest-complex of selected export-oriented spices and dry fruits and to develop control measures from the point of view of Quarantine.
8. *Rodent Control Unit:* During 1974, a Rodent Control Board was constituted. In 1980, an official level committee on Rodent Control substituted the Board, now restructured as an 'Expert Committee on Rodent Control'. Based on the recommendation of the Rodent Control Board earlier, a nucleus Rodent Control Unit has been set up in the Directorate.
9. *Pesticides Monitoring Unit:* This unit assesses the annual requirement of various pesticides, helps the Government of India in formulation of policies on various aspects of pesticides, helps in implementing 50 per cent pesticides distribution programme of the Government of India which involves quarterly distribution of indigenously manufactured pesticides to States/Union territories and collection of data on pesticides requirement, production, etc.
10. *Planning and Coordination Unit:* This unit coordinates the work relating to the Central Sector Scheme, Annual Plans,

---

* In 1968, Insecticide Act was passed to regulate the import, manufacture, sale, transport, distribution and use of insecticides with a view to prevent the hazard to humans and animals, and was enforced w.e.f. 1.8.1971.

TABLE 2.1: List of Pesticides provisionally registered under section 9 (3B) for data generation/large scale trial

S. No.	Name of Insecticides	As on 22.04.1994 Valid upto
1.	Iprodione	23.11.1994
2.	Isoprothiolane	24.11.1994
3.	Cyfluthrin	26.11.1994
4.	Methoprene	26.11.1994
5.	Propiconazole	24.05.1995
6.	Pretilachlor	28.05.1995
7.	Hexaconazole	09.08.1995
8.	Profenofos	10.11.1995
9.	Penconazole	11.11.1995
10.	Kasugamycin	21.11.1995
11.	Pyroquilon	30.01.1996
	* Neem Products	
12.	Neem triterpene emulsion containing Azadirachtin 0.15% w/w (1500 ppm) Margocide CK 20 EC	24.05.1995
13.	RD-9 Repellin (Neem oil 93 EC)	16.08.1995
14.	Azadirachtin 5% Conc. (Neemazol-F)	25.10.1995
15.	Neem oil based wsp containing Azadirachtin 0.03% (300 ppm) Godrej Achock	25.10.1995
16.	Neem oil emulsion containing 300 ppm Azadirachtin (Nimbecidine)	09.02.1996

* Commercialization of Neem-based and bio-pesticides is allowed to promote IPM technology

Source : Directorate of Plant Protection, Quarantine and Storage, Government of India, Faridabad (Haryana)

TABLE 2.2 : List of Pesticides not approved for use as on 28.02.1994

S. No.	Name of Pesticides
1.	Calcium Arsenate
2.	EPM
3.	Azinphos-Methyl
4.	Lead Arsenate
5.	Mevinphos (Phosdrin)
6.	2, 4, 5-T
7.	Carbophenothion
8.	Vamidothion
9.	Mephosfolan
10.	Azinphos ethyl
11.	Binapacryl
12.	Dicrotophos
13.	Thiocemeton/Disulfoton
14.	Fentin acetate
15.	Fentin hydroxide
16.	Chinomethionate (Morostan)
17.	Ammonium sulphamate
18.	Leptophos (Phosvel)

Source : Directorate of Plant Protection, Quarantine and Storage, Government of India, Faridabad (Haryana)

TABLE 2.3 : List of pesticides banned for use in India as on 28.02.1994

S. No.	Name of Pesticides
1.	Dibromochloropropane (DBCP)
2.	Endrin
3.	Pentachloronitrobenzene (PCNB)
4.	Pentachlorophenol (PCP)
5.	Toxaphene
6.	Ethyl parathion
7.	* Chlordane
8.	* Heptachlor
9.	* Aldrin
10.	Paraquat-di-methyl Sulphate
11.	Nitrofen
12.	Tetradifon
13.	PMA
14.	Nicotine sulphate

Source : Directorate of Plant Protection, Quarantine and Storage, Government of India, Faridabad (Haryana)

* These pesticides are under fresh review as per the directions of Hon'ble High Court of Rajasthan

TABLE 2.4: List of Pesticides restricted for use in India as on 31.08.1994

S. No.	Name of Pesticides	Type of restriction imposed
1.	Aluminium phosphide	It is to be sold only to Governments, Government undertakings/Organizations and to be used under the strict supervision of Government Experts or Pest Control Operators whose expertise is approved by the Plant Protection Adviser to the Government of India.
2.	B.H.C.	The use of BHC on vegetables, fruits, oilseeds crops and preservation of foodgrains is banned.
3.	Chlorbenzilate	Use in agriculture is banned. If required it can be imported by Government/Semi-Government organizations and prepared folbex strips for making it available to bee keepers for controlling mites of honeybees.

TABLE 2.4 *(Contd.)*

S. No.	Name of Pesticides	Type of restriction imposed
4.	Captafol	Shall be used only as seed dresser. Its use as foliar spray is banned.
5.	D.D.T.	Use of DDT in agriculture is banned. In very special circumstances warranting the use of DDT for Plant Protection, the state or Central Government may purchase it directly from M/s Hindustan Insecticides Ltd. to be used under expert Government supervison.
		Use of DDT for the Public health programme upto 10,000 MTs per annum except in case of any major outbreak is restricted.
6.	Dieldrin	Use of Dieldrin shall be restricted for locust control in desert areas by Plant Protection Adviser to the Government of India.
7.	Ethylene dibromide (EDB)	Use of EDB shall be restricted as a fumigant for foodgrains through Central Government, State Governments, Government undertakings, Government Organisations, such as Food Corporation of India, Central Warehousing Corporation, State Warehousing Corporation and Pest Control operators whose expertise is approved by Plant Protection Adviser to the Government of India.
8.	Methyl bromide	Methyl Bromide be used as a fumigant in the manner in which Aluminium Phosphide is used.
9.	Sodium cyanide	Use of Sodium Cyanide shall be restricted for fumigation of cotton bales by Plant Protection Adviser to the Government of India under expert supervision.
10.	Phenym mercury acetate (PMA)	Use of PMA is prohibited in India w.e.f. 1.01.1993 but it may continue to be produced in the country only for Export. Non-consumed stock as on 1.01.93 shall be used for export purpose only and shall be subject to the modification in the Certificate of Registration to this effect.
11.	Lindane	Use of Lindane formulation generating smoke for indoor use is prohibited in India, provided that Lindane may be continued to be produced in the country for use for the control of insect pests of field crops and subject to the modification in the Certificate of Registration to this effect.
12.	Methyl parathion	Use of Methyl parathion is permitted only on those crops where honey bees are not acting as pollinators and Methyl parathion may be continued to be

TABLE 2.4 *(Contd.)*

S. No.	Name of Pesticides	Type of restriction imposed
		produced in the country for use in other crops and subject to the modification in the Certificate of Registration to this effect.
13.	Nicotine sulphate	Use of Nicotine sulphate is prohibited in India provided that it may be continued to be produced in the country only for Export purpose and subject to the modification in the Certificate of Registration to this effect.

Source : Directorate of Plant Protection, Quarantine and Storage, Government of India, Faridabad (Haryana)

Quarterly Progress Review meetings, performance, budget etc.

11. *Documentation unit:* This unit receives, compiles and disseminates 'latest information on plant protection technology'.
12. *Plant Protection Training Institute:* This Institute conducts a ten-month Post Graduate Diploma course in Plant Protection and three-month course in (a) Advanced Plant Protection, (b) Pesticides formulation and (c) Pesticide residue analysis (Anonymous, 1986).

In recent years lot of expansion of the Directorate of Plant Protection Quarantine and Storage including the establishment of 39 Pesticides Testing Laboratories and 26 Central Integrated Pest Management Centres spread all over the country, has taken place (Anonymous, 1993).

Research Publications

The number of publications between 1938—50 were 1,500 (Roonwal, 1954, quoted by Pradhan and Jotwani, 1964), whereas those between 1953-62 were 3000—just double the number (Pradhan and Jotwani, 1964). This denotes the tremendous progress the science of Entomology made during the decade 1953—1962. Out of these, the highest number about 500 were on toxicology, about 350 on morphology and taxonomy, about 150 on biological control, about 200 each on bionomics and ecology, about 200 on insect physiology, genetics and cytology put together and the rest were on general entomological topics (Pradhan and Jotwani, 1964). Considerably large amount of work has been done on the taxonomy of termites, parasitic Hymenoptera, Thysanoptera and aphids. Work on ecology was restricted mainly to the effects of temperature and humidity on insect population. The monumental work of Y. Ramachandra Rao on Desert Locust was published in 1960 (Pradhan and Jotwani, 1964). A new biotic theory of the periodicity of locust cycles was propounded in 1961 (Pradhan, 1961). Insect Physiology studies have been going on, on digestive physiology, nutritional requirements, preparation of artificial diets for mass rearing of pests for experimental purposes, neurosecretory activity and endocrine glands and detoxification of insecticides. Possibility of the use of radioactive isotopes for the control of pests is also being explored. The work on these lines has been going on at the Indian Agricultural Research Institute as well as in State Departments of Agriculture and in certain universities.

Attention is now being paid to non-insect pests also. The work on mites was entrusted to systematic section of the Division of Entomology of the Indian Agricultural Research Institute (IARI), New Delhi and its control aspect is being dealt with by the section of Toxicology. Since 1964 work on phytophagous snails, their biology and control had been going on at the Indian Agricultural Research Institute and a section of Agricultural Malacology, was created. It continued to function upto the end of 1980. Now a section of Orinthology dealing with bird pests, also exists in the Division of Entomology, IARI, New Delhi. Some work on rodent control also is being undertaken by the Division of Entomology, IARI, New Delhi. Large scale work on rodents has been going on under a coordinated project of ICAR under the leadership of the Animal Ecologist at the Central Arid Zone Research Institute, Jodhpur. In a number of Agricultural Universities, Department of Zoology has come up after separation from the department of Entomology, to enable the scientists to deal with non-insect pests in a better way.

During the Second World War, Coyne was appointed by the Government of India as Director of Storage to take proper care of storage of imported foodgrains. He was succeeded by K.R. Sontakay. On the establishment of the Directorate of Plant Protection Quarantine and storage in 1946, the Storage Directorate was amalgamated with it. However, after a year it was again separated, merging with it the Directorate of Inspection of the Ministry of Food (Pruthi, 1969). With the increase in the import of foodgrains, the Directorate was very much expanded. Sontakay retired in 1958 and was succeded by S.V. Pingle. In 1969, his services were lent to the newly-created organization, the Food Corporation of India. After the movement of Pingle to Food Corporation, N.S. Agrawal succeeded him. Now there is a laboratory at Hapur (Uttar

Pradesh) conducting research on large scale grain storage problems.

Entomological work is also going on in Regional Research Stations of the Council of Scientific and Industrial Research. Indian Standard Institute, recently renamed as Bureau of Indian Standards has three standing sectional committees for (i) storage and marketing structure, (ii) pest control and (iii) apiary industry (Pradhan and Jotwani, 1964).

During the Second Five-Year Plan in the Division of Entomology, IARI, New Delhi, sections of Systematic Entomology, Insect Ecology and Toxicology, Insect Parasitology and Insect Physiology were established. Later sections dealing with the control of pests of sorghum, maize, pulses, cotton, sections of Insect Genetics, Agricultural Malacology, Locust Ecology, Insect Ecology, Ornithology, Plant Quarantine, Nematology, Radioactive isotopes, Insect Pathology and Seed Testing were also created. Further development took place as a result of which, the section of Nematology was raised to the status of an independent Division in 1966; the section of Radio-active isotopes was shifted to a newly-constructed Nuclear Research Laboratory having the radio-isotopes sections of all the major disciplines of agriculture, the Plant Quarantine section which was actually a part of Plant Introduction Division, became part of the newly created National Bureau of Plant Genetic Resources and shifted to a separate building in the Pusa campus. New Delhi, in August, 1976.

Entomological Seminars/Symposia

In January 1969, an International Seminar on Integrated Pest Control was organised jointly by ICAR, Entomological Society of India and a number of Pesticide firms under the Chairmanship of late Dr. S. Pradhan, Head of the Division of Entomology, IARI, New Delhi. By this time the publication of *Silent Spring* (1962) by Rachal Carson and reports of the development of resistance by certain insects against insecticides created a scare and suspicion amongst the public and even Entomologists. This Seminar restored confidence amongst the workers for it was clear that while the integrated control restricted the use of pesticides, it allowed the continued use of pesticides along with other methods of control e.g., integration of chemical, biological, ecological and mechanical methods. This further ensured reduction of hazards and pollution and reduction of selection pressure, leading to reduction of chances of development of resistance amongst the pests. Immediately after this Seminar, in another Seminar on “Sciences in Relation to India’s Agriculture in 1975", Late Dr. S. Pradhan pointed out that although potentialities of new approaches to pest control do present a bright picture, the need in the field of pest control in the country was urgent but this urgency was not being realised to the extent that the science of Entomology was not being given its due. He pointed out that the wheat revolution followed and not preceded the pest control revolution as was evident from the fact that Paul Muller, the discoverer of DDT, was awarded 'Nobel Prize' as early as in 1948, 22 years before Norman Borlogue got Nobel Prize (1970) for revolutionising wheat by the discovery of high yielding Mexican variety of wheat. In India, the wheat crop was practically free from pests hence does not need pesticidal umbrella whereas in America it has potent pests like the Hessian fly, it was only with the aid of entomological umbrella that it could succeed (Pradhan, 1969).

The grand success of this seminar resulted in restoring confidence amongst workers. In the last decade (1971-1980) symposia, seminars and workshops were held on (1) Pesticide Toxicology (New Delhi, 1975), (2) Recent Trends in Plant Protection (Allahabad, 1976), (3) Entomologists' role in Rural Development (Kalyani, 1976), (4) Plant Protection Research and Development (Coimbatore, 1976), (5) Insecticides of Plant Origin (New Delhi, 1978), (6) Pesticide Residues with reference to sampling technique (New Delhi, 1978), (7) Advances in Insect Taxonomy in India and the Oriental Region (Manali, 1979), (8) Integrated Pest Control Seminar (New Delhi, 1979), (9) High Altitude Entomology and Wild Life Ecology (Solan, 1979), (10) Pesticides for Rural Prosperity (New Delhi, 1979).

In recent years, a few important seminars and symposia held were (1) National Conference on Key Pests of Agricultural Crops (Kanpur, 1985), (2) Arid Zone Ecology (Jaipur, 1986), (3) Short Term and Long Term Hazards of Pesticides and Strategies for their safe use (New Delhi, 1986), (4) World Neem Conference (Bangalore, 1993), (5) First National Conference on Bee-keeping (Chandigarh, 1993), (6) National Conference on Eco-friendly Approaches in the Management of Pests, Diseases and Industrial affluents (Kanpur, 1993) and (7) Allelopathy in Sustainable Agriculture, Forestry and Environment (New Delhi, 1994).

As regards the non-insect pests, the problem of rodents was discussed in a symposium at Kanpur in 1971 and at Jodhpur in 1978. Economic Ornithology was discussed in a workshop at Bharatpur in 1979 and Acarology at Bangalore in the same year. A symposium on non-insect pests and predators was held at New Delhi in 1984.

The second International Conference on Apiculture in Tropical Climates was held at New Delhi from February 29 to March 4, 1980, sponsored jointly by the Indian Council of Agricultural Research, Khadi and Village Industries Commission, Department of Science and Technology, Indian National Science Academy from India and the International Bee Research Association, U.K. In all, eighty-two papers were read at the conference dealing with practically all aspects of Bee Keeping. This conference focussed the attention towards usefulness of insects like honey-bees. In addition to the production of honey, bees have a much greater role to play in pollinating about 80 per cent of the fruit and vegetable crops as also a number of seed and fodder crops depend for their pollination and fertilization on insects, amongst whom bees are the most effective pollinators.

To commemorate this Conference on apiculture, Indian Posts and Telegraph Department issued a special one rupee postage stamp on February 29, 1980, the day the conference was inaugurated at New Delhi.

Today Entomology is a full-fledged subject in all the Agricultural Colleges and Universities of India and Pakistan.

Publication of Entomological Books

Some of the comparatively more important Entomology books published by the Indian authors in the last 5 decades are:

1. *Economic Entomology for South India* by Shri T.V. Ramakrishna Ayyar, published in 1940; a revised edition was published in 1963.
2. *The Desert Locust in India* by Rao Bahadur Y. Ramachandra Rao, published in 1960.
3. *Entomology in India,* edited by Dr. N.C. Pant, published in 1964.
4. *General Entomology* by Dr. M.S. Mani published in 1968, latest edition published in 1982.
5. *Insect Pests of Crops* by Dr. S. Pradhan published in 1969, revised edition published in 1983.
6. *Text Book of Agricultural Entomology* by Dr. H.S. Pruthi, published in 1969.
7. *General and Applied Entomology* by Drs. K.K. Nayar, T.N. Ananthakrishnan and B.V. David, published in 1976, latest edition published in 1986.
8. *Agricultural Entomology and Pest Control* by Dr. S. Pradhan published in 1983.

Some other books on Entomology published in recent years are:

A Monograph on Thysanoptera by Dr. T. N. Ananthakrishnan was published in 1969, three books dealing with High Altitude Insects by Dr. M.S. Mani were published in 1962, 1963 and 1968, and a collection of Summer Institute Lectures on Insect Anatomy and Physiology edited by Drs. N.C. Pant and S. Ghai was published in 1975 the revised edition of which was published in 1981. Two other books: *Principles and Concepts of Integrated Pest Management* edited by Drs. R. A. Agarwal, G.P. Gupta, Prem Kishore and Dinesh Chandra and *Recent Advances in Entomology* edited by Drs. Y. K. Mathur, A. K. Bhattacharya, N. D. Pandey, K. D. Upadhyaya and J.P. Srivastava were published in 1983 and 1987 respectively. Pesticides: Their ecological impact in developing countries edited by Drs. G. S. Dhaliwal and Balwinder Singh was published in 1993, Bee-keeping Proc. 1st National Conference held at Chandigarh in 1993, edited by Drs. N. P. Goyal and R. K. Sharma published in 1993, Biological Pest, Suppression by Drs. R. D. Gautam published in 1994 and Hand Book of Crop Pest Control by Drs. K. K. Marwaha, K. H. Siddiqui and J.P. Singh published in 1994.

Although there is "No yardstick to measure the exact gains of better crop protection, the fact that occasional outbreak of the pests and diseases and epidemics have been contained and the green revolution has been sustained, is adequate proof of achievements of research in Plant Protection" (Randhawa, 1979).

Entomology has not yet taken the Very First Place

In spite of all these developments Entomology has not yet been given its very first place. So far there has been only one Entomologist as Director of IARI, out of all the Director Generals of ICAR so far, none has been an Entomologist. There are a number of Deputy Director Generals dealing with various agricultural disciplines but Entomology is unrepresented. The entire Plant Protection i.e., Entomology, Mycology, Plant Pathology and Nematology are being dealt with by two Asstt. Director Generals.

Mian Afzal Husain, in his Presidential address to the section of Entomology at the Indian Science Congress in 1938 said that "there cannot be any doubt whatsoever that among sciences of economic value in India, Entomology must take the very first place in importance", but he said that" it occupies at present the very last". Even after fifty-seven years of this historic statement, Entomology still does not occupy the very first place. Let us see when the dream of Mian Afzal Husain is fulfilled.

3

PHYLUM ARTHROPODA—ITS CLASSES

A Comparative Study

The word 'Arthropoda' is of Greek origin. Gr. *arthros* = joint and Gr. *podas* = foot. Thus Arthropods are joint-footed animals—a character common to all the members of the phylum.

Arthropoda is the largest phylum of the animal kingdom. (Snodgress, 1952). This phylum contains more than 891,000 species (Storer and Usinger, 1965). According to Miller and Haub (1960), there are about 1,000,000 species of animals existing in the world, of which 800,000 animals i.e., 4/5th belong to phylum Arthropoda. Frost (1942) also states "that Arthropoda are about 80 per cent of all the species of the animal kingdom (Fig. 2).

Arthropods occur upto the height of over 6080 metres on land and upto a depth of about 59,728 metres in water. Different arthropod species are adapted for living in air, on land, in soil and in fresh, brakish or salt water (Storer and Usinger, 1965).

Many arthropods are economically important: crabs, lobsters, prawns and shrimps form a part of human diet; small-sized crustaceans are staple food of fishes; some insects and spiders are eaten by many land vertebrates including man. Many insects are pests of crops, stored grains, stored food, household goods and clothing etc. Some insects and archnids are in one way or the other responsible for certain diseases of crops, domestic animals and human beings.

Although this phylum includes animals, widely differing in structure,

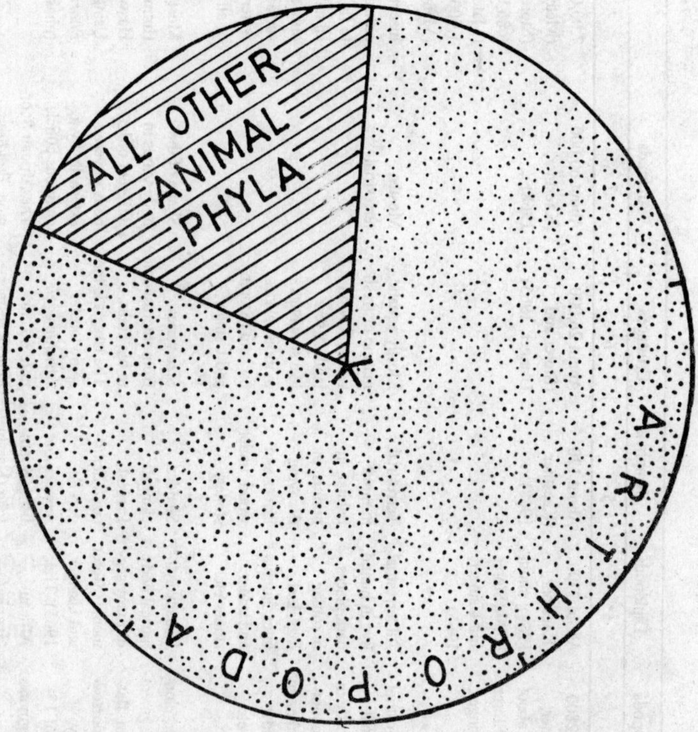

FIG. 2 : Numerical Strength of Arthropoda in relation to the rest of the animal phyla (After Frost, 1942)

TABLE 3.1: Comparative Study of the Major Arthropod Classes

		Onychophora	Chilopoda	Diplopoda	Symphyla	Crustacea	Archnida	Insecta
	1	2	3	4	5	6	7	8
1.	No. of species	70 (Kaestner, 1968; Storer and Usinger, 1965)	About 2800 (Kaestner, 1968); about 3000 (Storer and Usinger, 1965).	About 7500 (Kaestner, 1968); about 8000 (Storer and Usinger, 1965).	About 120 (Kaestner, 1968),	About 26,000 (Storer and Usinger, 1965)	About 60,000 (Kaestner, 1968)	7,00,000 species described (Metcalf, 1940); 8,00,000 (Storer and Usinger, 1965); 6,86,000 species described (Matcalf and Flint, 1979); 1,500,000 species described (Mani, 1982).
2.	Habitat	All terrestrial	All terrestrial nocturnal, live mostly in warm countries, hiding by day under stones etc.	All terrestrial, live in humid, dark places in rotting logs, leaf litter, humus and under stones etc.	Terrestrial, in moist habitat, in humus, in decaying leaves and wood, under stones,	Mostly aquatic, marine or fresh water, few on land. A few like sow bugs in moist places on land (Storer and Usinger, 1965)	Mostly, terrestrial	Insects are the only arthropods which are able to fly. Found in practically all habitats-terrestrial or aerial, only a few groups live on the surface of or in fresh/brakish water, in soils, on/in plants of all kinds, on/in animals.
3.	Size and shape	Caterpillar like animal, cylindrical in form (Parker and Haswell, 1951) The largest	Slender, elongate, segmented form, flattened dorsoventrally. Body of 15-181 segments.	Usually cylindrical body of many segments (Storer and Usinger, 1965). The African *Graphido-*	Slender, elongate; Largest-sized *Scuitigerella immaculata* being 8 mm long (Kaestner,	Size varies from microscopic to big crabs. The largest is the spiny lobster *Jasus huegeli* upto 60 cm long (Kaestner,	Shape of the body varies in different order; body covered externally by a chitinous cuticle. The African Scorpion, *Pandinus,*	Most insects are 2-40 mm long (Imms, 1965; Parker and Haswell, 1951; Storer and Usinger, 1965). Walking stick *Pharnacia serratipes* (phasmid) is 33 cm long.

TABLE 3.1 *(Contd.)*

1	2	3	4	5	6	7	8
	sp. *Peripatus quatus* can grow to a length of 15 cm. (Kaestner, 1968)	Some tropical species are 15-20 cm long (Storer and Usinger, 1965). The South American *Scolopendra gigantea* can grow to a length of 26.5 cm (Kaestner, 1968).	*streptus gigas* is 28 cm long (Kaestner, 1968).	1968).	1968). Shape of body varies, body covered externally by a chitinous cuticle which becomes thickened in regions where no movement is required (Parker and Haswell, 1951).	*imperator*, may reach a length of 18 cm (Kaestner, 1968).	
4. Body Divisions	No division, head continuous with body. Only feature of segmentation is short stumpy legs (Fig. 3)	Two divisions; head and long body (Fig.4)	Head, short thorax and long abdomen (Storer and Usinger, 1965) (Fig. 5)	Two divisions; head and body; body segments 15-22, the dorsal and ventral segments do not correspond (Fig. 6)	Two divisions; usually cephalothorax (head + thorax) and abdomen, usually with a a carapace (Fig. 7)	Two divisions; cephalothorax and abdomen (Fig. 8)	Three distinct divisions, head, thorax and abdomen (Fig. 9)
(i) Head	No distinct head	Head distinct	Head distinct	Head distinct	Usually head fused with thorax.	Not distinct	Head distinct
(ii) Thorax	Not separate	Thorax not distinct	Thorax short, four simple	Not distinct	Fused with head	Fused with head	Three segmented.

TABLE 3.1 *(Contd.)*

1	2	3	4	5	6	7	8
			segments				
(iii) Abdomen	Not separate.	Abdomen not distinct.	Abdomen with 20-100 double somites.	Made up of 15 segments.	Abdominal somites usually distinct with telson at the end.	Abdomen distinct	Eleven or fewer segmented abdomen with terminal part modified as genitalia.
5. Appendages							
(i) Antenna	One pair, short, made up of a number of short rings bearing minute spines.	One pair, long, with 12 or more joints.	One pair, short and very similar to the legs.	Single pair, long filiform.	Two pairs, one pair antennae and one pair antennule; antennule = antennae of insects.	Antennae absent.	One pair of antennae.
(ii) Mouth Parts	Mouth rimmed by a fleshy fold containing 2 small horny jaws and 2 blunt oral papillae (Storer and Usinger, 1965; Hickman, 1966).	One pair of mandibles and 2 pairs of maxillae (Storer and Usinger, 1965).	Mandibles one pair and maxillae one pair (Parker and Haswell, 1951).	Two pairs of maxillae and one pair of mandibles (Parker and Haswell, 1951).	One pair of mandible, 2 pairs of maxillae and 2 pairs of maxillipedes (Storer and Usinger, 1965).	One pair of claw-like chelicerae, one pair of pedi-plapi—short, leg—like with enlarged bases forming "maxillae" for squeezing and chewing the food (Storer and Usinger, 1965). No mandibles	One pair of mandibles, two pairs of maxillae and a labium. Mouth parts adapted for chewing, sucking and/or lapping.

TABLE 3.1 *(Contd.)*

1	2	3	4	5	6	7	8
						(Hickman, 1966).	
(iii) Legs	Many pairs-one pair per somite, un-jointed ending in 2 claws (Storer and Usinger, 1957; Parker and Haswell, 1951).	Numerous-one pair per somite. Ist somite bears a pair of 4-jointed poison claws and on each of the other except last two, are small paired, 7 jointed walking legs (Storer and Usinger, 1965).	Numerous, the four thoracic segments have one pair of legs per somite; each abdominal somite has two pairs of legs (Hickman, 1966).	One pair of posterior un-jointed appendages and 12 pairs of jointed legs in adults, each having two claws (Storer and Usinger 1965).	At least 5 pairs of walking legs, one pair per somite, or legs variously modified, some usually biramous e.g., in prawn. (Storer and Usinger, 1965).	Four pairs on cephalothorax in adult stage; in some larval forms only 3 pairs (Storer and Usinger, 1965).	Three pairs of jointed thoracic legs, size of different pairs may vary according to the requirement (Fig. 10). In some there are only two pairs of functional walking legs. The anterior most (forelegs) may be modified for other purposes e.g., in *Mantis* and *Gryllotalpa* (Figs. 20, 24).
6. Wings	Absent	Absent	Absent	Absent	Absent	Absent	Usually two pairs; in some only one pair or absent.
7. Digestive system	Alimentary canal has buccal cavity, a thick-walled pharynx, narrow oesophagus, mesentron, a narrow rectum	Straight with 2 or 3 pairs of salivary glands (Storer and Usinger, 1965).	As in Chilopoda (Storer and Usinger, 1965).	The foregut enlarges into extendible midgut in segment 4 or 5; there are no caeca and at about segment 10 enters the ectodermal	Straight and tubular, 2 large digestive glands opening in midgut separately.	Digestive system with mouth, oesophagus, stomach, intestine, digestive glands, rectum and anus. Poison glands, spinning glands in some.	Digestive tract of fore, mid and hind gut, mouth with salivary gland (Imms, 1965; Storer and Usinger, 1965).

TABLE 3.1 *(Contd.)*

	1	2	3	4	5	6	7	8
		leading to anal aperture in the last body segment. Two long narrow salivary glands.			hindgut (Kaestner, 1968).			
8.	Sense organs	Antennae, sensory hairs and eyes (Storer and Usinger, 1965).	Antennae and eyes present; in *Lithobius* sp. organs of Tomosvary may have auditory function (Kaestner, 1968).	Antennae and eyes present (Storer and Usinger, 1965). Organs of Tomosvary studied in *Glomerius* sp. appear to be olfactary in function (Kaestner, 1968).	Antennae sensory; setae and organs of Tomosvary present. Usually no eyes (Kaestner, 1968).	Sensory setae and tactile hairs for sense of touch; chemoreceptors in hairs on the antennules tips of antennae, mouth parts and chelae; statocyst for equilibrium and orientation of gravity.	Eyes present, pedipalpi and pectines tactile; response to sound uncertain; in spiders, minute "lyriform organs" are olfactory (Storer and Usinger, 1965), Trichobothium present and sensitive to air currents and vibrations (Kaestner, 1968).	Eyes present, chemoreceptors for smell on the antennae, for taste about the mouth are various tactile hairs, some with means for sound production and reception. Tympanic membrane sensitive to sound vibrations.
9.	Eyes	One pair of simple eyes (Storer and Usinger, 1965).	Single or cluster of simple eyes. (Parker and Haswell, 1951).	Two clumps of many simple eyes. (Storer and Usinger, 1965).	Usually absent	One pair of compound eyes on stalks. (Parker and Haswell, 1951).	Simple ocelli (Storer and Usinger, 1965).	A pair of compound eyes, simple eyes also present (Imms, 1965; Parker and Haswell, 1951.)
10.	Nervous System	Consists of a pair of cere-	Brain and sub-oesophageal	Similar to chilopoda,	The brain relatively	Presence of supra-oesophageal and	Bilobed ganglia, ventral ganglionic	Brain—a bilobed mass of nerve matter, situated in the head,

TABLE 3.1 *(Contd.)*

1	2	3	4	5	6	7	8
	bral ganglia (brain), several anterior nerves, two circumpharyngeal connectives and a pair of ventral nerve cords without true ganglia, joined together at the posterior end behind the anal aperture (Storer and Usinger, 1965; Parker and Haswell, 1951).	ganglia present. The ganglia are present throughout the body which forms a long chain with medially fused metameric pairs of ganglia. The pairs of ganglia are separated by paired connectives.	except that the paired ganglia are not outwardly connected by connectives.	large. Protocerebrum lacking, globuli and an optic mass, receives nerve fibres from organs of Tomosvary.	sub-oesopha-geal ganglia and nerve cord.	mass and a pair of connectives. Paired ganglia occur in abdomen of young spider not in adult (Storer and Usinger, 1965).	two parallel nerve cords, ventrally situated, and jointed at intervals by ganglia (Miller and Haub, 1960; Parker and Haswell, 1951).
11. Circulatory system	Body cavity a haemocoele, circulatory system consists of a	Heart extends the length of body surrounded by pericardium with a	As in Chilopoda (Storer and Usinger, 1965).	The heart extends from metamere 3 to 13, behind the posterior border	Open (lacunar) blood system	Heart is slender, dorsally placed with 3 pairs of ostia (Storer and Usinger, 1965).	Heart slender with an anterior aorta, no capillaries or veins, body, spaces a haemocoele. (Imms, 1965; Storer and Usinger, 1965).

TABLE 3.1 *(Contd.)*

1	2	3	4	5	6	7	8
	tubular heart with ostia.	pair of ostia and of lateral arteries in each somite (Storer and Usinger, 1965).		of segment 5, it has 8 pairs of ostia (Kaestner, 1968).			
12. Respiratory system	Delicate un-branched or rarely bran-ched trach-eal tubes. Groups of these open in little depre-ssions of the integument- the tracheal pits, external openings of which are known as stigmata (Parker and Haswell, 1951).	In some somites there are spir-acles joining a fused system of tracheal tubes (Storer and Usinger, 1965).	Many separate tracheae each a tufted pouch connected to a spiracle in front of leg (Storer and Usinger, 1965).	Single pair of branching tracheae, the two external apertures of which are stipulated in the head (Parker and Haswell, 1951).	Respiration by gills, or by body surface. Pseudotr-achea on pleopods of some land isopods (Storer and Usinger, 1965).	Respiration by book lungs which are peculiar to arachnids, opening on two sides of epigynum. Trachea if present restricted to the abdomen (Storer and Usinger 1965).	Respiration by tracheae that carry oxygen from paired spiracles on sides of thorax and abdomen, directly to the tissues. Some with tracheal or blood gills (Imms, 1965).
13. Excretory system	Excretory organs	Two long Malpi-ghian tubules	Two long Mal-pighian tubules	One pair of Malpighian	Excretion by 1 or 2 pairs of green glands.	Excretion by paired Malpighian tubules	Excretion by 2 to many fine Malpighian tubules attached to

TABLE 3.1 *(Contd.)*

1	2	3	4	5	6	7	8
	are pairs of nephridia-like structures situated in pairs on the lateral compartment of the body cavity and open on the lower surface of the legs at their bases. Segmentally arranged nephridia (Parker and Haswell, 1951).	for excretion (Storer and Usinger, 1965).	for excretion.	tubules and the maxillary nephridia (Kaestner, 1968).	No Malpighian tubules (Storer and Usinger, 1965).	and one or two pairs of coxal glands in the floors of the cephalothorax that empty by ducts between the legs. These are considered homologous with the green glands of crustacea (Storer and Usinger, 1965).	anterior end of hind gut (Imms, 1965).
14. Reproductive system.	Sexes separate, reproductive organs are paired but with a single external opening. Reproductive system includes paired	Sexes separate, each having one dorsal gonad and paired accessory glands connected to a ventral genital opening near the posterior end (Storer,	Sexes separate, gonad is single and ventral with a duct opening on the 3rd somite. Male has modified appendages (gonopods) on the 7th somite.	Sexes separate, Paired ovaries, a pair of oviducts opening through a median atrium between the coxae of the 4th pair of legs. Testes are in the same position, fused for	Sexes separate except in some parasitic forms, sex openings paired (Storer and Usinger, 1965).	Sexes separate, female larger than male. Male has two testes joined by coiled vasa efferentia to a single seminal vesicle leading to the genital opening. Female with 2 ovaries. Oviduct joining the vagina, a lateral seminal recept-	Sexes separate, Female larger than male, gonads of multiple tubules with one median duct posteriorly; fertilization internal; ova with much yolk and protective shells (Imms, 1965; Parker and Haswell, 1951).

TABLE 3.1 *(Contd.)*

1	2	3	4	5	6	7	8
	testes seminal vesicles, vasa deferentia, accessory glands and one ejaculatory duct. (Storer and Usinger, 1965); Parker and Haswell, 1951). Reproductive ducts ciliated.	and Usinger, 1965).	Fertilization is internal and each egg is enclosed in a covering of regurgitated food (Storer and Usinger, 1965).	some distance, a pair of complex vasa deferentia opening through a median atrium on sternite 4 (Keastner, 1968).		acles (Storer and Usinger, 1965).	
15. Sex Opening	One, at the end of the body (Storer and Usinger, 1965).	One, at the end of the body (Storer and Usinger, 1965).	One, on the third somite near the head. (Storer and Usinger, 1965).	One, near the head on the third body somite (Storer and Usinger, 1965).	Two, at the hind part of the thorax (Storer and Usinger, 1965).	One, anteriorly in the mid line on 2nd somite of abdomen and covered by a plate or epigynum (Storer and Usinger, 1965).	One, at the end of abdomen. (Imms, 1965).
16. Development	Direct, most species are viviparous and part of each oviduct is specialized as a uterus where the	Direct, lays eggs or viviparous.	Direct, lays eggs.	Direct, lays eggs.	Usually with larval stages. Parthenogenesis seen in some forms. (Storer and Usinger, 1965).	Direct, except in mites and ticks (Storer and Usinger, 1965).	(1) Direct, metamorphosis incomplete, pupal stage wanting; several nymphal instars. (2) Indirect, metamorphosis usually complete, pupal stage present between larval and adult stages. (3) Parthenogenesis in some.

TABLE 3.1 *(Contd.)*

	1	2	3	4	5	6	7	8
		embryos develop. In some forms a placenta like trophoblast develops. A large female may produce 30—40 young per year (Storer and Usinger, 1965).						
17.	Behavioural, protective, defensive adaptations and devices	Negatively phototropic, when disturbed ejects mucous several times from adhesive glands of oral papillae. This adhesive is used as defence against enemies.	Negatively phototropic, injects poison by mouth—a defensive, protective adaptation	Secrete poisonous chemicals through stink glands, a pair in each segment with paired openings on each segment. They move their antennae continuously while moving, touching the ground; the sensory organs are at the	Negatively phototropic. They move through soil and response to change in humidity. Secretion of sticky silks by the spinnerets when touched—a protective adaptation.	Inflict injury by chela. Legs modified according to requirement for aquatic/terrestrial life.	All species project their pedipalpi; spiders inflict injury by their mouth while scorpions inflict injury through their stings situated at the posterior end of the body. Colour adaptations to environment in some cases. Some spiders mimic ants, move unnoticed and keep on predating on them.	Both nocturnal and diurnal; senses well developed particularly chemoreceptors on antenna and maxilla are well-developed. They respond to change in temperature and humidity also. Eyes are well developed to perceive vision. Defensive mechanisms resemblance of one species to another, resemblance to surroundings; death feint; presence of poisonous hairs, secretion of bad-smelling liquid and construction of protective structures. Exploitation of exoskeleton maximum; insects are the only arthro-

TABLE 3.1 *(Contd.)*

	1	2	3	4	5	6	7	8
				tip of antennae.				pods having functional wings. Legs modified variously for digging, jumping, holding the prey.
18.	Economic importance	No economic importance. However, they are of academic interest as they demonstrate a connecting link between Annelida and Arthropoda.	Predatory, prey upon insects, earthworms, lizards and sometimes even mice.	Generally feed on dead plant/ animal materials. They also feed on living plant materials hence are crop pests, damaging under ground parts of certain plants of economic importance. Some secrete phenol (Eisner, Hurst and Meinwald, 1963a) or produce hydrogen cyanide (Eisner, and Hurst, 1963 b) Only some species are predatory secreting hydrocyanic	Practically no economic importance except that sometimes they are found damaging roots and root hairs of green house plants, thus, retarding the growth of plants. However, these are of great academic interest because of the symphylan theory of insect origin.	Serve as an important item of human diet. Prawns, lobsters, crabs etc. are relished by man. Small crustaceans serve as food for many fishes of economic importance. Certain crabs are pests of crops e.g., *Birgus latro*, the coconut crab is a serious pest of coconut palm; certain species of crabs are pests of aquatic crops like paddy and *Trapa.* Hermit crabs are predatory and feed upon snails (Srivastava, 1968).	Arachnids are pests, predators and disease carriers. Scorpions are predatory, feed upon flies, cockroaches, crickets, grass-hoppers, beetles etc. They inject poison from their poison glands through their stings. Mites are pests of crops but certain species are predatory also; ticks as a whole are parasitic on man or animals. Pathogenic organisms are transmitted by various mites/ticks to man/ domesticated animals. Spiders are predatory and kill the prey by injecting poison through their chelicerae and	Insects are of great economic importance. Several species damage different parts of plants; damage is caused by biting and chewing/sucking plant sap, by rolling leaves, by making galls, by making cases out of plant parts, by boring/tunnelling in different plant parts, by feeding on stored food grains and stored products, by damaging household materials, clothes etc. Some species are poisonous, still others are vectors of animal (including man) and plant diseases. There are also insect species valuable to mankind. Some species serve as dyes used in beverages and cakes; some are of medicinal importance e.g., *Lytta* sp., the blister beetle gives the 'canthardin' and honey bee's extract the 'apis' used in certain

TABLE 3.1 *(Contd.)*

1	2	3	4	5	6	7	8
			acid from their stink glands (Srivastava and Srivastava, 1967).			suck out the prey. Spiders spin webs extensively on a number of plants and retard their growth. Spiderbite is generally harmless but bites of genus *Lactodectus* may sometimes be fatal. This genus is found in Canada, South America and Hawaii islands.	human diseases as medicines (Frost, 1942). Some species act as pollinators e.g., bees, certain beetles and flies etc. Certain fruit trees, berries and plants of cucurbitaceae family are almost wholly dependent upon insects for pollination, otherwise fruit-setting may not take place. There are crops which give better yield by cross-pollination through insects. Certain species supply us useful, valuable products like honey, silk and lac. Insects also form a part of human diet in some countries including some parts of India. Some species are used in pest management as parasites/predators of pests. Some like dung rollers, thysanurans and collembolans act as scavengers.

they do agree in certain fundamental characters: arthropods are bilaterally symmetrical, metamerically segmented, possess hardened chitinous exoskeleton and paired jointed appendages exhibiting functional modifications in different regions of the body. In addition, there is a central nervous system with ganglionated ventral nerve cord, a dorsal heart with pericardium and paired ostia, body cavity is not a true coelom but a haemocolele which is in free communication with the circulatory system. Muscles are composed almost entirely of striated muscle fibres and ciliated epithelium is generally absent. No animals other than arthropods exhibit the above combination of characters (Imms, 1965). A comparative account of the major arthropod classes is given in Table 3.1

From the comparative study of the different classes of phylum Arthropoda, it is evident that insects share many characters with other arthropods, but the other arthropods have not been able to exploit these to the extent insects have done. The presence of functional wings—a very important character which insects do not share with other arthropods has given insects distinct advantage over other arthropods in their struggle for existence and dominance.

Why numerically insects dominate over other arthropods and the animal kingdom as a whole, has been dealt with in detail in Chapter 4.

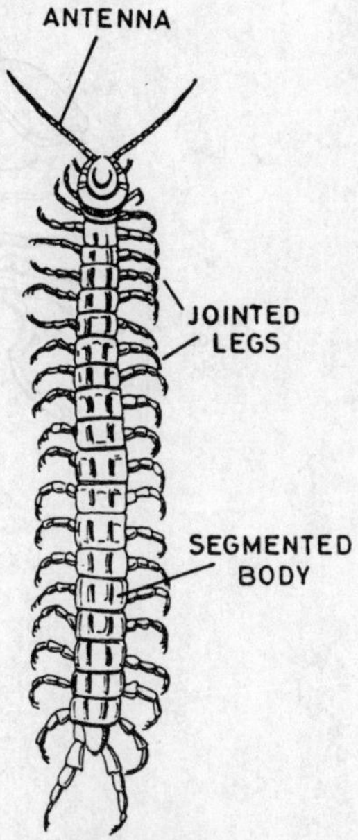

FIG. 3: *Peripatus*—Oncychophora
(After Parkar and Haswell, 1951)

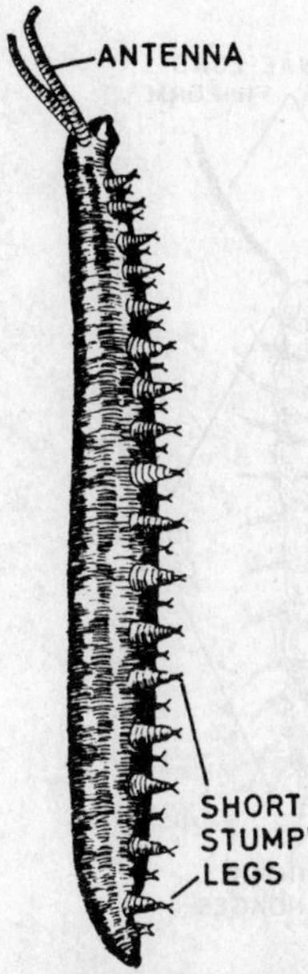

FIG. 4: *Scolopendra*—Chilopoda
(After Parkar and Haswell, 1951)

FIG. 5: *Orthomorpha* sp.—Diplopoda

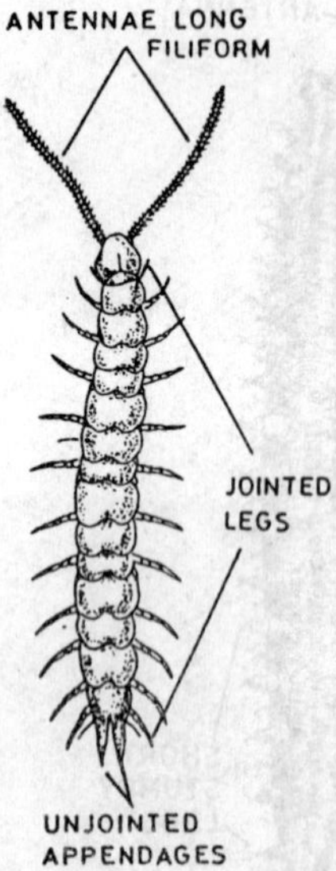

FIG. 6: *Scutigerella immaculata*
—Synphyla
(After Kaestner, 1968)

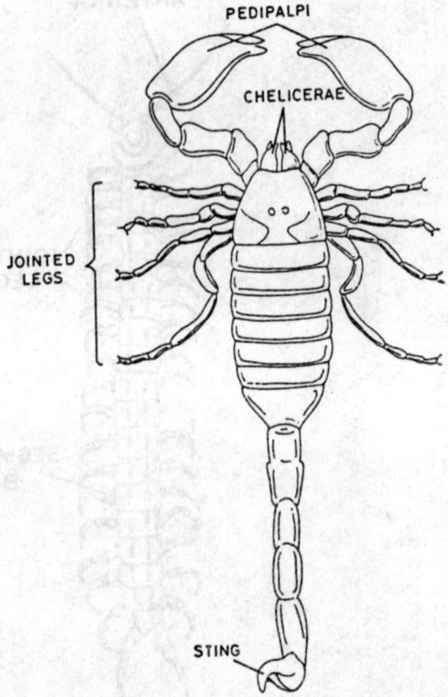

FIG. 8: Scorpion—Arachnida
(After Parkar and Haswell, 1951)

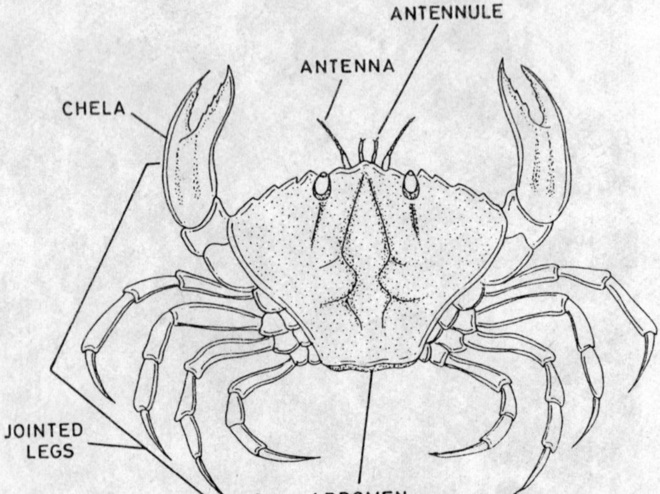

FIG. 7: A land crab—Crustacea
(After Parkar and Haswell, 1951)

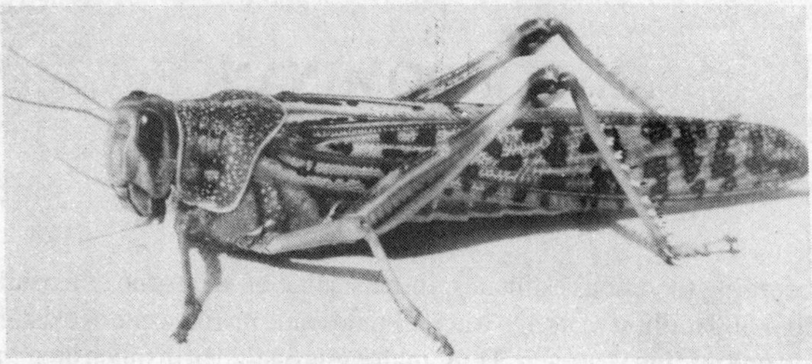

FIG. 9: *Schistocerca gregaria*—Insecta

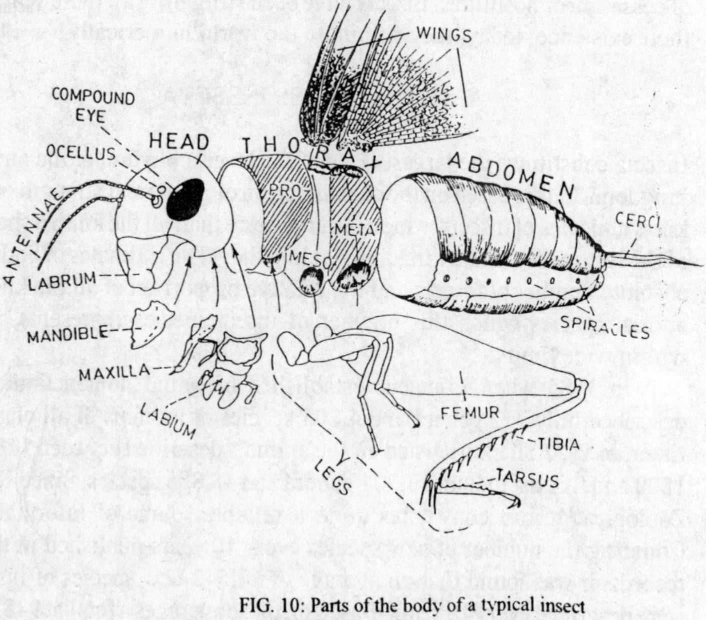

FIG. 10: Parts of the body of a typical insect
(After Metcalf and Flint, 1979)

4

INSECT DOMINANCE

According to various estimates, insects came on the scene of existence 250—500 million years ago whereas man came on the scene of existence only one million years ago. Thus, man came on the earth much later than the insects. Prior to man's appearance, insects must have been living peacefully as man was not there to compete with them. Ever since man came into existence a regular warfare has been going on without any sign of cessation of hostilities. Insects have been struggling all these years for their existence; today, they dominate the world numerically.

A. SIZE OF CLASS INSECTA

Insecta constitutes the largest class of the largest phylum of the animal kingdom. This is based on the consideration of numerical strength of the known species of insects, which is much more than all the known species of the rest of the animal kingdom put together. The estimates of both the absolute number of insects and the relative proportion of all the known animal species which this number of insect species represents, vary within wide limits.

In 1785 when Linnaeus established binomial nomenclature he described only 312 genera and 4,203 species of animals of all classes. Later, in 1902 Sherborn listed all the animals described between 1785 to 1800 and this list includes 3,204 genera and 48,855 species. Since 1850, Zoological record constitutes quite a reliable source of information. Counting the number of new species every 10 years published in these records, it was found that an average of 10,542 new species of insects were described every 10 years. Based on these averages a total of 948,780 species of insects and a total of 652,770 species of animals other than

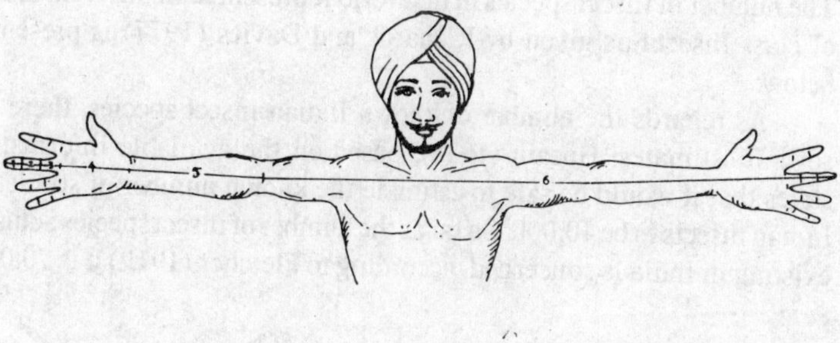

FIG. 11: Position of insects in the animal kingdom (After Ayyar, 1940).

insects were described between 1850 and 1940. This appears to be the most reliable estimate available. Approximately 700,000 species of insects have been described (Z.P. Metcalf, 1940) but this number may not be representing even one fifth of the species existing today (Imms, 1965). A single order Coleoptera alone has 330,000 species out of which a single family Curculionidae has more than 60,000 species while the family Chrysomelidae is the close second numbering over 26,000 species (Imms, 1965). Metcalf and Flint (1979) state that more than 75 per cent of all the animals hitherto found and named belong to phylum Arthropoda, and about 90 per cent of these are 'true insects'. According to them about 686,000 different kinds of insects have been discovered, properly named and described and thousands of new species are being found every year indicating that we are far from having reached a full knowledge of class Hexapoda. According to Ayyar (1940) taking the distance between the finger tips of our outstretched right and left arms across over chest to represent the size of the whole animal kingdom, class Insecta occupies a distance stretching from the elbow of one arm across the chest and the two shoulders right upto the tip of the large finger of the other arm (Fig. 11), thus showing its enormous size when compared with all the other animal groups. According to Gosard (1909) the number of species that are probably living in all parts of the earth range from 2,500,000 to 10,000,000. Metcalf and Flint (1979) state that according to the average of two other estimates the total number of animal species works out to be 2,500,000 of which the number of insect species alone is estimated to be 1,500,000. Mani (1982) also states that the total number of insect species described from the world exceeds 1,500,000 representing nearly 90 per cent of the animal kingdom.

B. ORDERS OF CLASS INSECTA

The number of insect species in the world represented in different orders of class Insecta as given by Richards and Davies (1977) is presented below:

As regards the number of known Indian insect species, there are several estimates. Husain (1938), based on the available information states that it would be safe to estimate the known number of species of Indian insects to be 40,000. So far as the number of insect species actually existing in India is concerned, according to Fletcher (1919) it is 70,000.

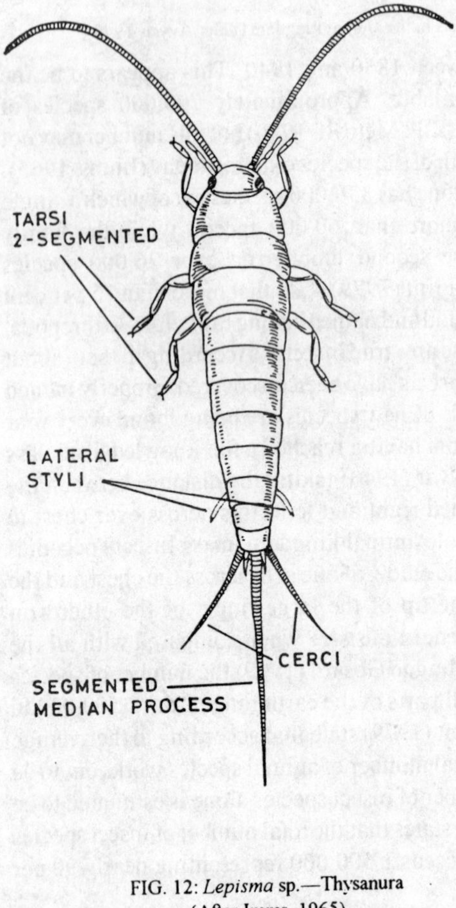

FIG. 12: *Lepisma* sp.—Thysanura
(After Imms, 1965)

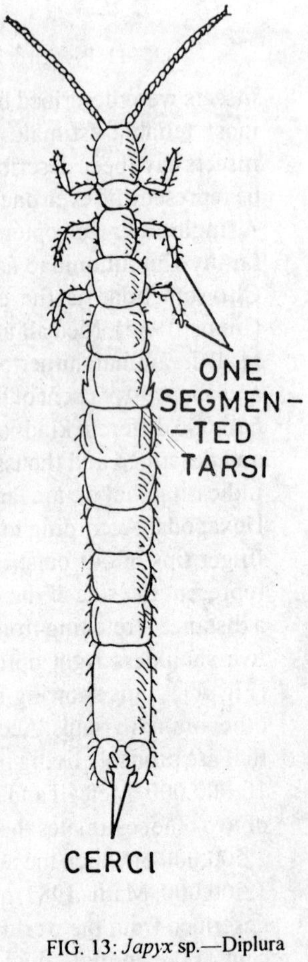

FIG. 13: *Japyx* sp.—Diplura
(After Pruthi, 1969)

Class Insecta is divided into two sub-classes Apterygota and Pterygota. Apterygota is divided into 4 orders. Pterygota has two divisions, (1) Exopterygota divided into 16 orders and (2) Endopterygota divided into 9 orders. General characters of all the 29 orders of class Insecta and the number of species belonging to each order are given below:

Subclass I, Apterygota. Insects without wings. Wingless condition presumed to be primitive. Metamorphosis slight or absent. Adult with one or more parts of pregenital abdominal appendages. Mandibles of adults, usually articulating with the head capsule at a single point.

1. *Thysanura* (500 species). Insects belonging to this order are commonly known as silverfish, 'bristle tails'. The general characters of the order are: ectognathous mouthparts adapted for biting, antennae many-segmented, but only the basal segment provided with muscles. Compound eyes present or absent. Tarsi with 2-4 segments. Abdomen 11-segmented with a variable number of lateral, styliform, pregenital appendages, a pair of many-segmented cerci and ending in a segmented median process (Fig. 12) Malpighian tubules present.
2. *Diplura* (600 species). Insects belonging to this order are known as campodeides, japygids. Distinguishing characters of the order are: entognathous mouthparts. Antennae many-segmented. Compound eyes and ocelli absent. Tarsi I-segmented. Abdomen with lateral styliform appendages on most or all of the pregenital segments and ending in paired cerci of variable form (Fig. 13). Terminal median filament absent. Malpighian tubules vestigial or absent.
3. *Protura* (more than 200 species). Insects belonging to this order are commonly known as 'telson-tails'. Insects of this order are minute insects with entognathous piercing mouthparts, antennae and compound eyes wanting. Abdomen composed of 11 segments and a well-developed telson, first three segments each with a pair of small appendages (Fig. 14) Malpighian tubules represented by papillae.
4. *Collembola* (more than 1500 species). Insects of this order are also known as spring-tails. The general characters of the order are: mouthparts entognathous, principally adapted for biting, antennae usually 4-segmented, compound eyes absent. Abdomen 6-segmented, usually with three pairs of

appendages, i.e., a ventral tube on segment I, a minute retinaculum on III, and a forked springing organ furcula on IV (Fig. 15). There are no Malpighian tubules.

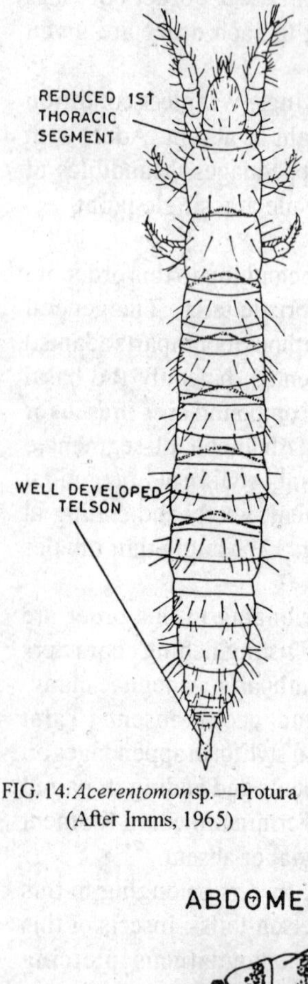

FIG. 14: *Acerentomon* sp.—Protura
(After Imms, 1965)

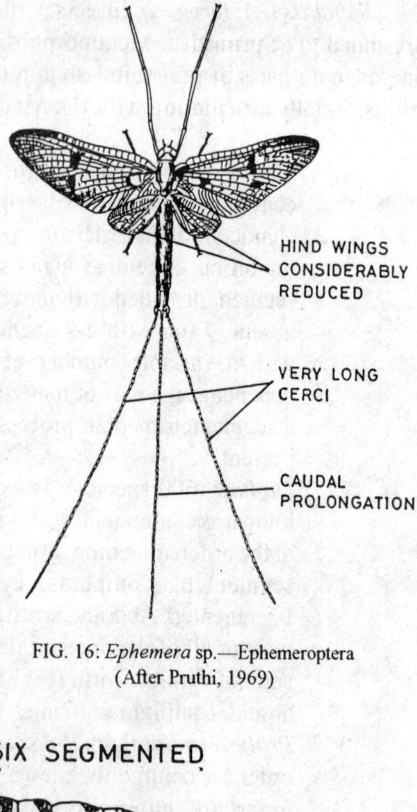

FIG. 16: *Ephemera* sp.—Ephemeroptera
(After Pruthi, 1969)

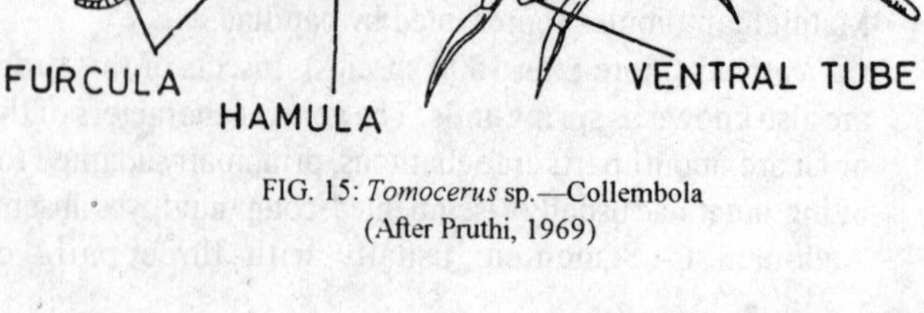

FIG. 15: *Tomocerus* sp.—Collembola
(After Pruthi, 1969)

Subclass II. Pterygota. Insects with wings or secondarily apterous. Metamorphosis varied, rarely slight or wanting. Pregenital abdominal appendages absent in adults. Mandibles in adults articulating with head capsule at two points, unless highly modified.

Division 1. Exopterygota (Hemimetabola)

Metamorphosis simple, sometimes slight, pupal instars rarely present. Wings develop externally. Immature stages usually resemble adults in structure and habits including feeding habits.

5. *Ephemeroptera* (2,000 species). Insects belonging to this order are commonly known as mayflies. These are soft-bodied insects with short setaceous antennae. Wings membranous, held vertically upwards when at rest, the hind pair considerably reduced, 'intercalary' veins and numerous cross-veins present. Abdomen terminated by very long cerci either with or without a similar median caudal prolongation (Fig. 16).
6. *Odonata* (4,500 species). Insects belonging to this order are commonly known as Damselflies, Dragonflies. These insects are predacious with biting mouthparts. Wings with complex reticulation of small cross-veins and usually a conspicuous stigma. Eyes very large and prominent, antennae very short and filiform (Fig. 17) Abdomen elongate, often extremely slender. Male accessory genital armature developed on 2nd and 3rd abdominal sterna.
7. *Plecoptera* (more than 1700 species). Insects belonging to this order are commonly known as stoneflies and are soft-bodied insects having setaceous antennae, ligula4-lobed. Wings membranous, held flat over the back in repose, hind pair usually the larger, with well developed anal lobes. Venation variable, often considerably specialized, vein M 2-branched. Tarsi 3-segmented (Fig. 18) Abdomen usually terminated by long multi-articulate cerci. Ovipositor wanting.
8. *Grylloblattodea* (16 species). The grylloblattids are apterous insects, with eyes reduced or absent and no ocelli. Antennae moderately long and filiform. Mouthparts mandibulate. Legs approximately similar to each other, tarsi 5-segmented. Female with well-developed ovipositor, either 7th or 8th abdominal sternum enlarged to form subgenital plate, cerci

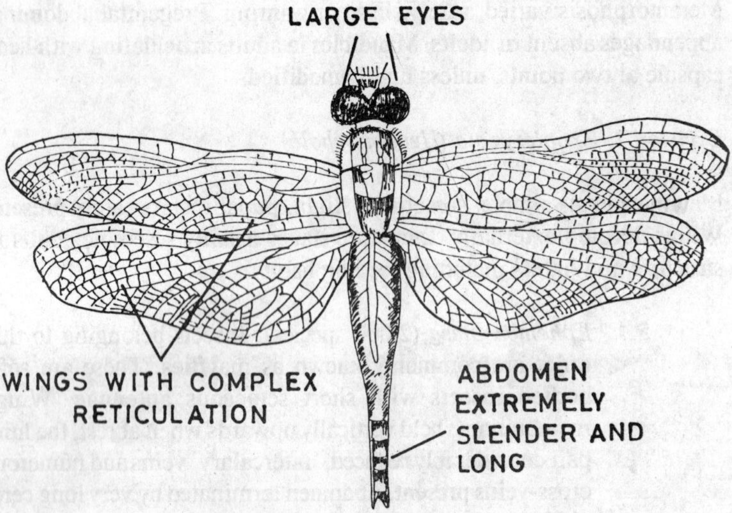

FIG. 17: *Acisoma* sp.—Odonata
(After Pruthi, 1969)

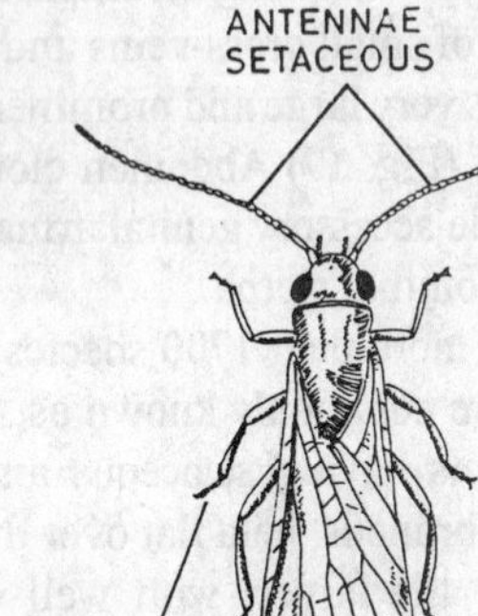

FIG. 18: *Perla* sp.—Plecoptera
(After Pruthi, 1969)

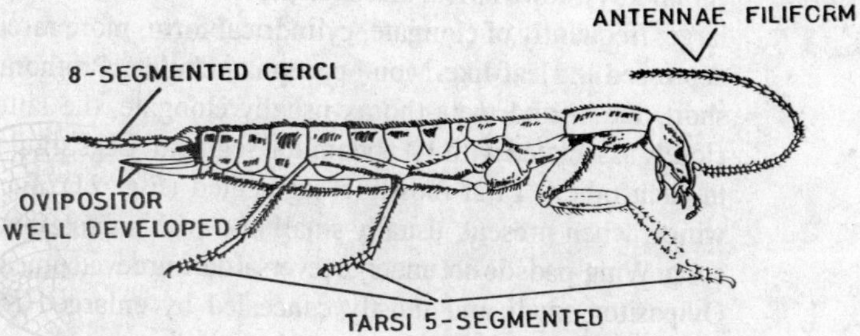

FIG. 19: *Grylloblatta* sp.—Grylloblattodea (After Pruthi, 1969)

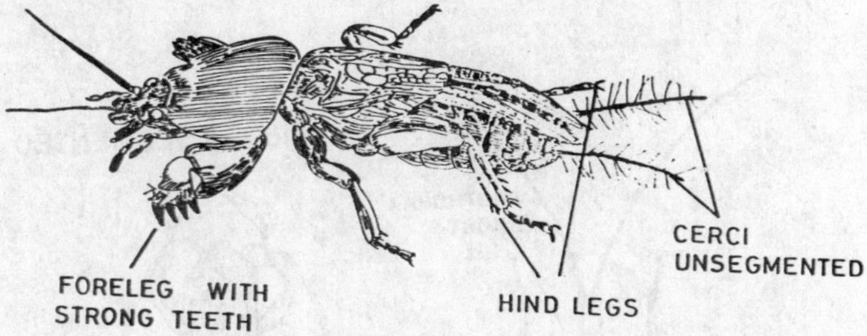

FIG. 20: *Gryllotapa* sp.—Orthoptera (After Imms, 1965)

long, 8-segmented (Fig. 19) Male genitalia asymmetrical.

9. *Orthoptera* (more than 17,000 species). Insects belonging to this order are commonly known as locusts, grasshoppers, crickets (Fig. 20). The general characters of the order are: mouthparts mandibulate. Prothorax large. Hind legs usually enlarged and modified for jumping, tarsi 3 or 4 segmented. Fore wings forming more or less thickened tegmina with submarginal costal vein. Wing pads of nymph undergo reversal during development. Female generally with well-developed ovipositer, not concealed by 7th or 8th abdominal sterna. Male external genitalia symmetrical, concealed at rest by enlarged 9th abdominal sternum which may or not bear a pair of styles. Cerci usually short and almost invariably unsegmented. Specialized auditory and stridulatory organs frequently developed.

10. *Phasmida* (2,500 species). Insects belonging to this order are commonly known as leaf and stick insects. These insects are large, frequently of elongate, cylindrical form, more rarely depressed and leaf-like. Mouthparts mandibulate. Prothorax short, meso- and meta-thorax usually elongate, the latter closely associated with 1st abdominal segment. Legs similar to each other; Tarsi mostly 5-segmented (Fig. 21). Fore wings, when present, usually small and with submarginal costa. Wing-pads do not undergo reversal during development. Ovipositor small and mostly concealed by enlarged 8th abdominal sternum. Male external genitalia variable and asymmetrical, concealed by 9th abdominal segment. Cerci short, unsegmented. Specialised auditory and stridulatory organs absent.

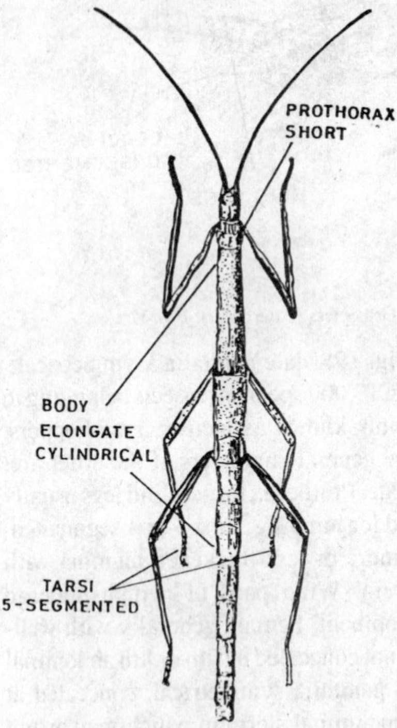

FIG. 21: *Carausius* sp.—Phasmida
(After Imms, 1965)

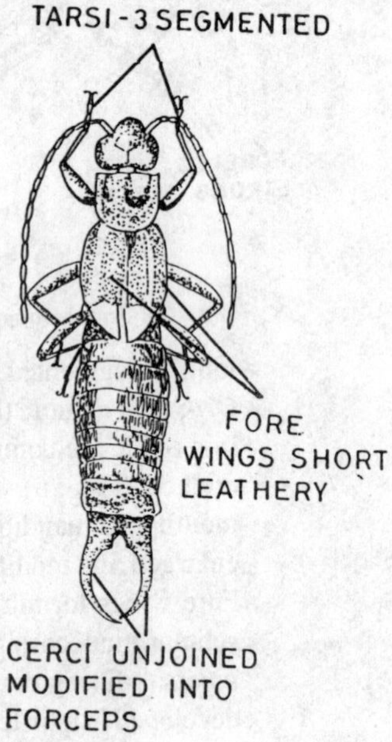

FIG. 22: *Forficula* sp.—Dermaptera
(After Imms, 1965)

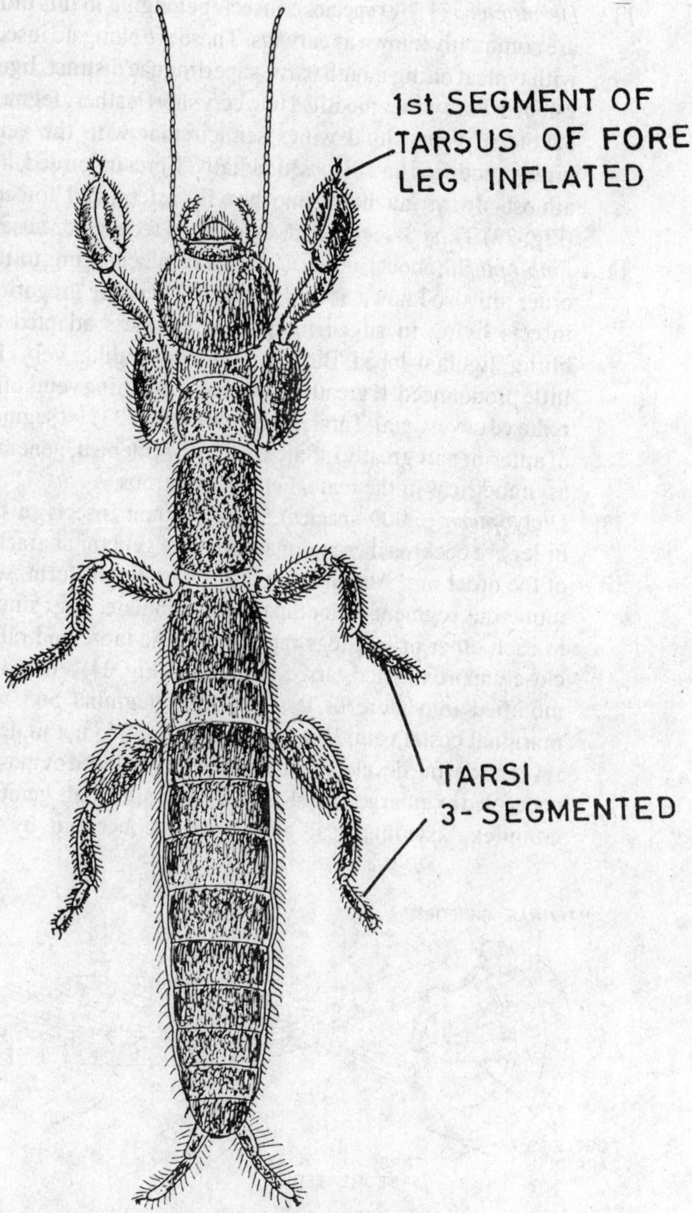

FIG. 23: *Embia major*—Embioptera
(After Imms, 1965)

11. *Dermaptera* (1,200 species). Insects belonging to this order are commonly known as earwigs. These are elongate insects with typical biting mouth parts, superlinguae distinct, ligula 2-lobed. Fore wings modified into very short leathery tegmina devoid of veins, hind wings semicircular with the veins highly modified and disposed radially. Cerci unjointed, and almost always modified into heavily sclerotized forceps. (Fig. 22) Tarsi 3-segmented. Ovipositor reduced or absent.
12. *Embioptera* (about 300 species). Insects belonging to this order are also known as webspinners. These are gregarious insects living in silken tunnels. Mouthparts adapted for biting, ligula 4-lobed. Both pairs of wings alike: veins but little pronounced, R greatly thickened, remaining veins often reduced or vestigial. Tarsi 3-segmented (Fig. 23) 1st segment of anterior pair greatly inflated. Cerci 2-semented, generally asymmetrical in the male. Females apterous.
13. *Dictyoptera* (6,000 species). The common insects of this order are cockroaches and mantids. The general characters of the order are: Antennae almost invariably filiform, with numerous segments. Mouthparts mandibulate. Legs similar to each other or fore legs raptorial, coxae large and rather closely approximated, tarsi-5 segmented (Fig. 24). Forewings modified into more or less thickened tegmina and with marginal costal vein. Wing-pads of nymph do not undergo reversal during development. Female with reduced ovipositor concealed by enlarged 7th abdominal sternum. Male genitalia complex, asymmetrical and largely concealed by 9th

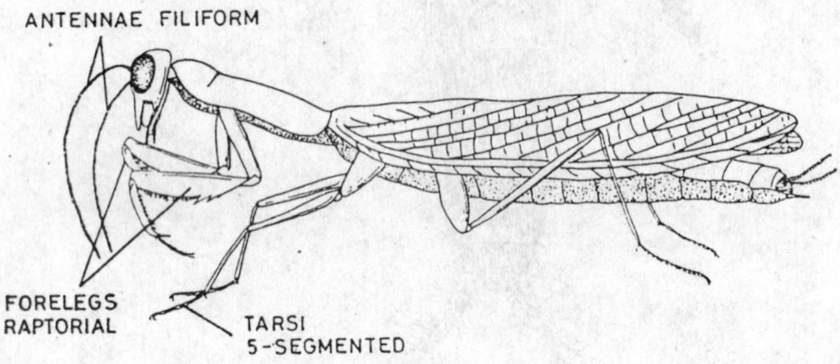

FIG. 24: *Mantis* sp.—Dictyoptera

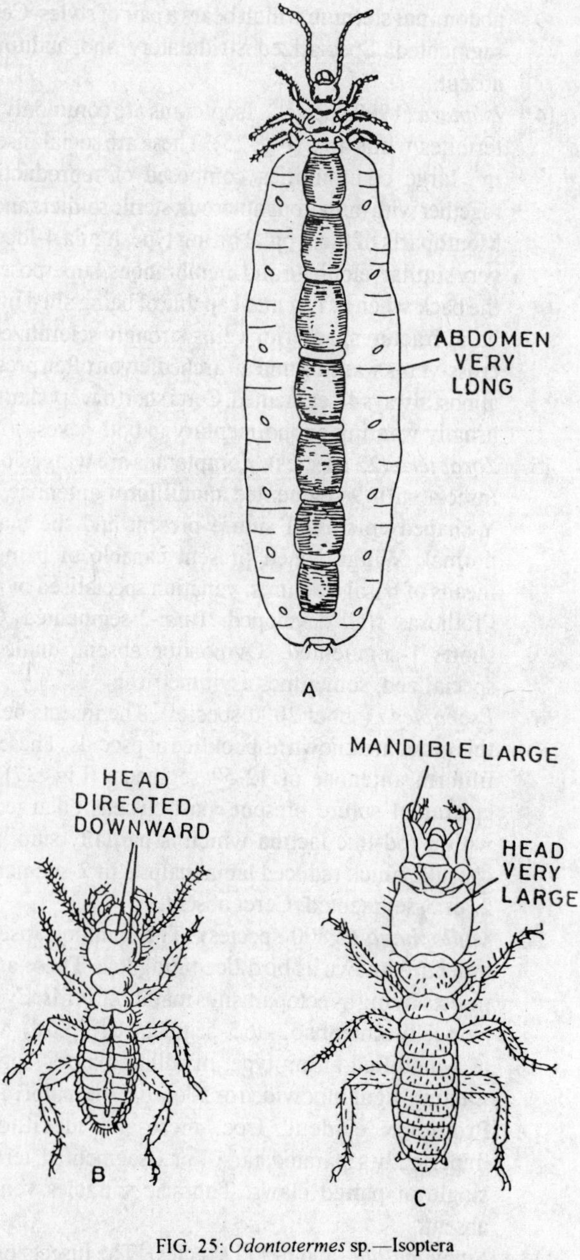

FIG. 25: *Odontotermes* sp.—Isoptera
(a) Queen (b) Worker (c) Soldier (After Pruthi, 1969)

abdominal sternum which bears a pair of styles. Cerci many-segmented. Specialized stridulatory and auditory organs absent.

14. *Isoptera* (1900 species). Isopterans are commonly known as termites/white ants (Fig. 25). These are social insects living in large communities composed of reproductive forms together with numerous apterous, sterile soldiers and workers. Mouthparts of the typical biting type, ligula 4-lobed. Wings very similar, elongate and membranous, superposed flat over the back when at rest and capable of being shed by means of basal fractures, anterior veins strongly sclerotized, regular cross-veins wanting and an archedictyon often present. Tarsi almost always 4-segmented. Cerci short or very short; genitalia usually wanting or rudimentary in both sexes.

15. *Zoraptera* (22 species). Zorapterans are winged or apterous insects with 9-segmented moniliform antennae (Fig. 26). Y-shaped epicranial suture present and the maxillae are normal. Wings, when present capable of being shed by means of basal fractures, venation specialised by reduction. Prothorax well developed. Tarsi-2-segmented. Cerci very short, 1-segmented. Ovipositor absent, male genitalia specialized, sometimes asymmetrical.

16. *Psocoptera* (about 2000 species). The insects belonging to this order are known as booklice or psocids. These have long filiform antennae of 12-50 segments (Fig. 27) Y-shaped epicranial suture present, post-clypeus enlarged. Maxilla with a rod-like lacinia which is partially sunk into head-capsule, much reduced labial palps 1 or 2-segmented. Tarsi 2- or 3-segmented. Cerci absent.

17. *Mallophaga* (2,800 species). The common insects of this order are known as bird lice/biting lice. These are apterous insects living as ectoparasites mainly of birds. Eyes reduced. No ocelli. Antennae 3- to 5- segmented (Fig. 28). Mouthparts of a modified biting type, maxillary palpi 4-segmented or wanting, ligula undivided or 2-lobed, labial palpi rudimentary. Prothorax evident, free, meso-and metathorax often imperfectly separated, tarsi 1 or 2-segmented, terminated by single or paired claws. Thoracic spiracles ventral. Cerci absent.

18. *Siphunculata* (about 300 species). The insects of this order are referred to as sucking lice. Members of this order are

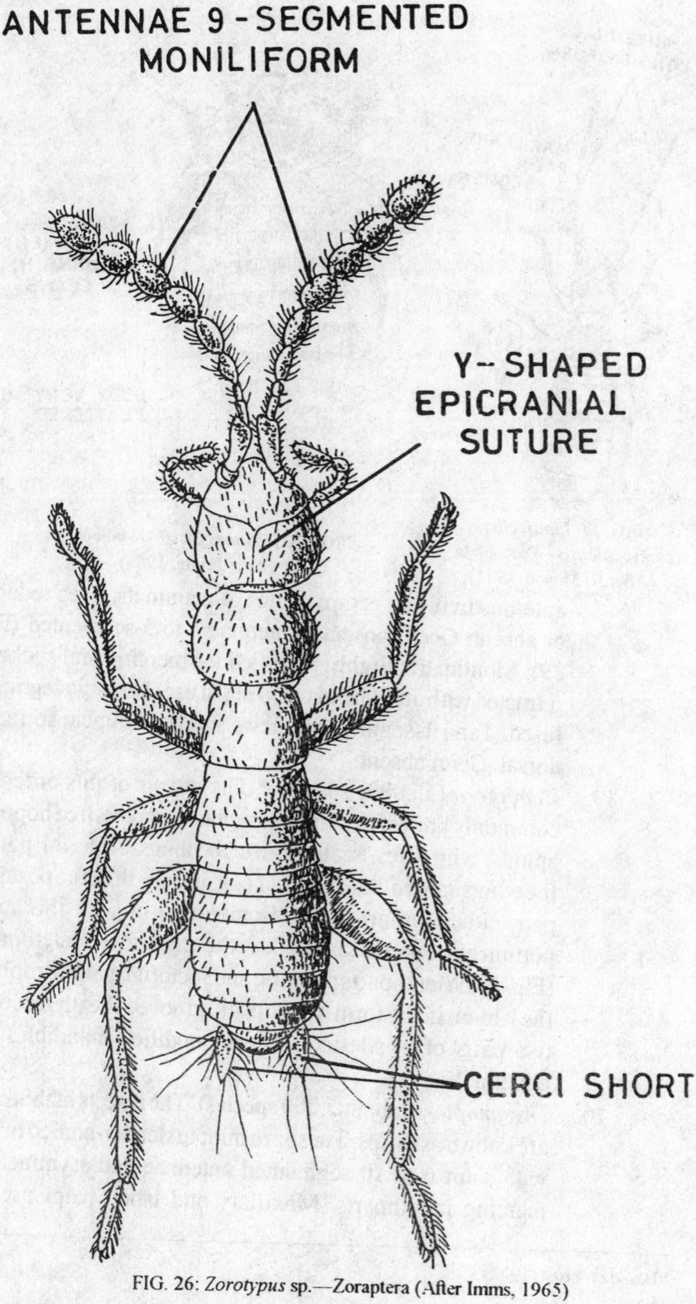

FIG. 26: *Zorotypus* sp.—Zoraptera (After Imms, 1965)

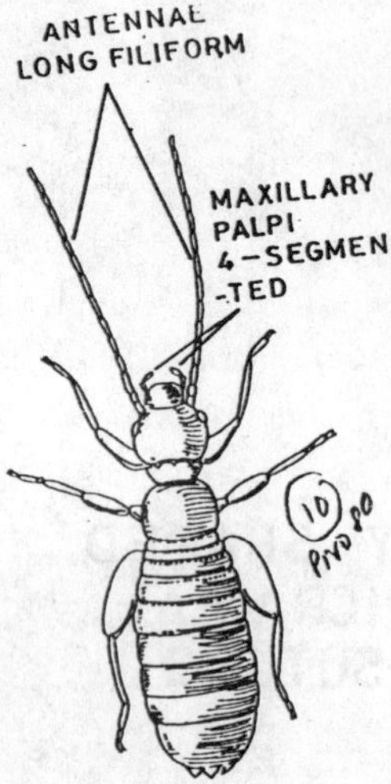

FIG. 27: *Liposcellis transvaalensis*—Psocoptera (After R. Menon, 1941)

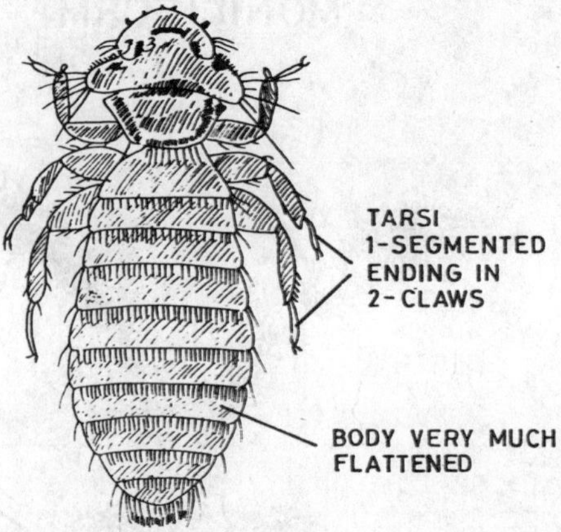

FIG. 28: *Osborniella* sp.—Mallophaga (After Pruthi, 1969)

apterous living as ectoparasites of mammals. Eyes reduced or absent. Ocelli absent. Antennae 3 to 5-segmented (Fig. 29). Mouthparts highly modified for piercing and sucking, retracted within the head when not in use. Thoracic segments fused. Tarsi 1-segmented, claws single. Thoracic spiracles dorsal. Cerci absent.

19. *Hemiptera* (55,000) species.* The insects of this order are commonly known as plant bugs, plant hoppers, tree hoppers. aphids, whiteflies, scale insects, bedbugs. Anterior pair of the wings most often of harder consistency than the posterior pair, either uniformly so (Homoptera) or with the apical portion more membranous than the remainder (Heteroptera) (Fig. 30). Mouthparts piercing and suctorial, palpi atrophied, the labium in the form of a dorsally grooved sheath receiving two pairs of bristle-like stylets (modified mandibles and maxillae).

20. *Thysanoptera* (about 5,000 species). The insects of this order are known as thrips. These are minute slender-bodied insects with short 6 to 10 segmented antennae and asymmetrical piercing mouthparts. Maxillary and labial palpi present.

---

* Based on Frost (1942).

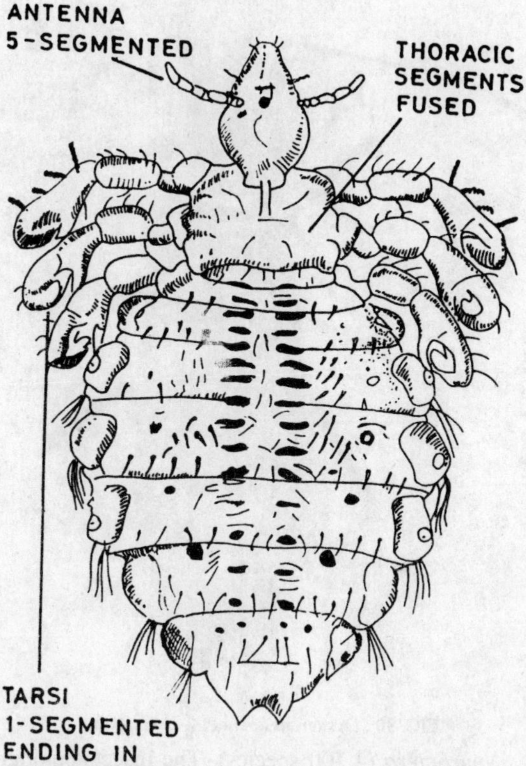

FIG. 29: *Haematopinus* sp.—Siphunculata (After Pruthi, 1969)

Prothorax well developed. Wings when present very narrow with greatly reduced venation and long marginal setae. Tarsi 1 or 2 segmented (Fig. 31) each with a terminal protrusible vesicle. Cerci absent.

Division 2. Endopterygota (Holometabola).

Metamorphosis complete, pupal instar always present. Wings develop internally. Immature stages, differ from adults in structure and habits including feeding habits.

FIG. 30 : *Dysdercus koenegii*—Hemiptera

21. *Neuroptera* (4,700 species). The insects belonging to this order are commonly known as lace wings, ant loins etc. These are small to rather large soft-bodied insects with usually elongate antennae. Mouthparts adapted for biting. Ligula undivided or bilobed or often atrophied. Two pairs of very similar membranous wings, generally disposed in a roof-like manner over the abdomen when at rest. Venation primitive but with many accessory veins, costal veinlets numerous. Abdomen without cerci (Fig. 32).
22. *Coleoptera* (3,33,000 species). The insects belonging to this order are commonly known as beetles, weevils etc. These are minute to large insects whose fore wings are modified into horny or leathery elytra which almost always meet to form a straight mid-dorsal suture (Fig. 33). Hind wings membranous, folded beneath the elytra, or often reduced or wanting.

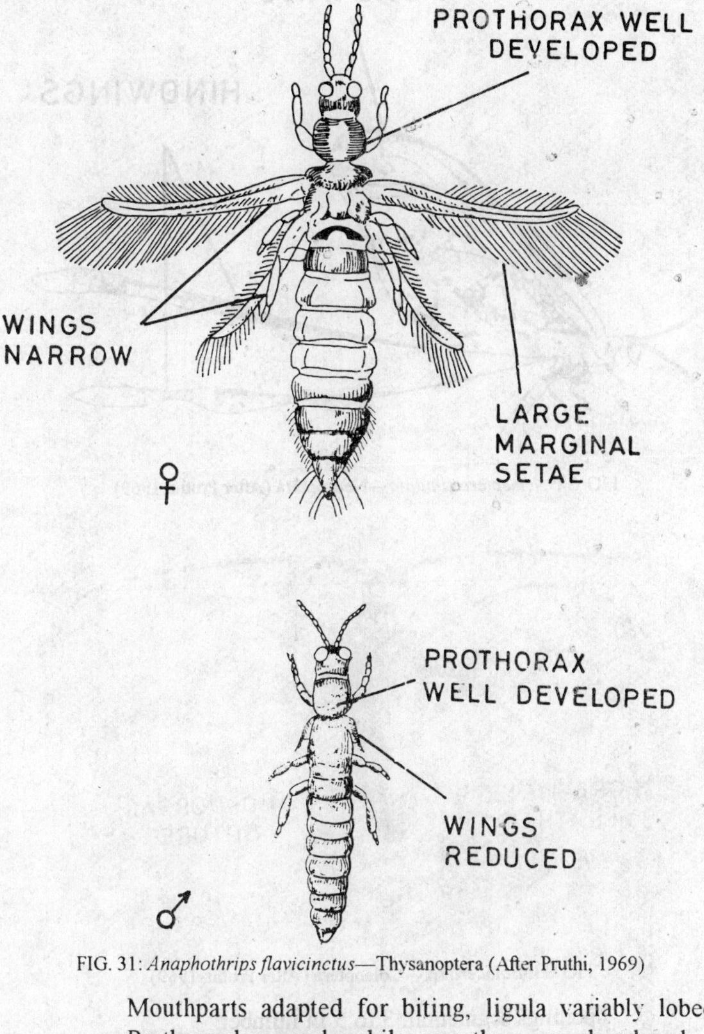

FIG. 31: *Anaphothrips flavicinctus*—Thysanoptera (After Pruthi, 1969)

Mouthparts adapted for biting, ligula variably lobed. Prothorax large and mobile, mesothorax much reduced.

23. *Strepsiptera* (370 species). The insects of this order are known as stylopids, are small and endoparasitic. Males free-living, mouthparts of a degenerate biting type, antennae conspicuous and flabellate (Fig.34), fore wings reduced to small club-like structures, hind wings large and fan-shaped, trochanters absent, metathorax greatly developed. Female normally remaining in host, enclosed in a puparium, sexual

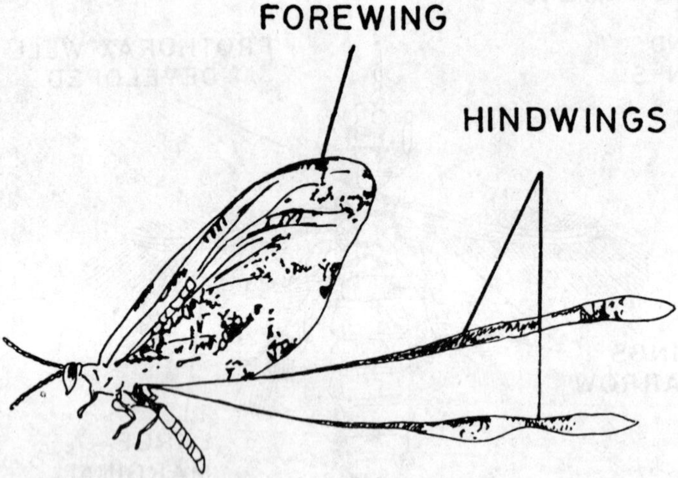

FIG. 32: *Nemeptera sinulata*—Neuroptera (After Pruthi, 1969)

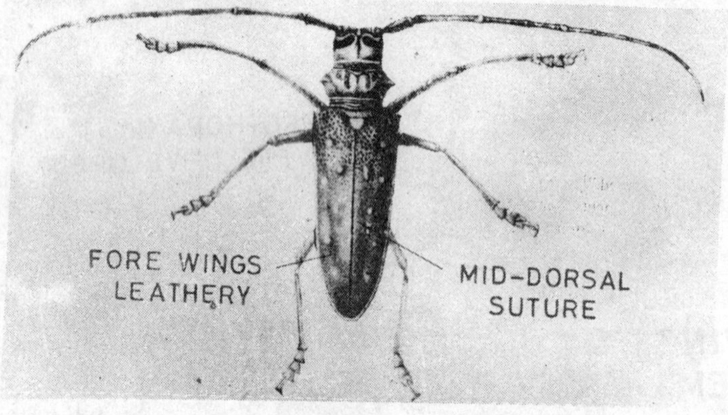

FIG. 33: *Batocera* sp.—Coleoptera (After Pruthi, 1969)

openings segmental, 3 to 5 in number.

24. *Mecoptera* (less than 400 species). The insects of this order are known as scorpion flies. These are slender, moderate or small-sized, usually carnivorous with elongate, filiform antennae (Fig. 35). Head usually produced into a vertically deflected rostrum, with biting mouthparts, ligula wanting, legs long and slender. Wings similar and membranous, carried longitudinally and horizontally in repose. Venation primitive,

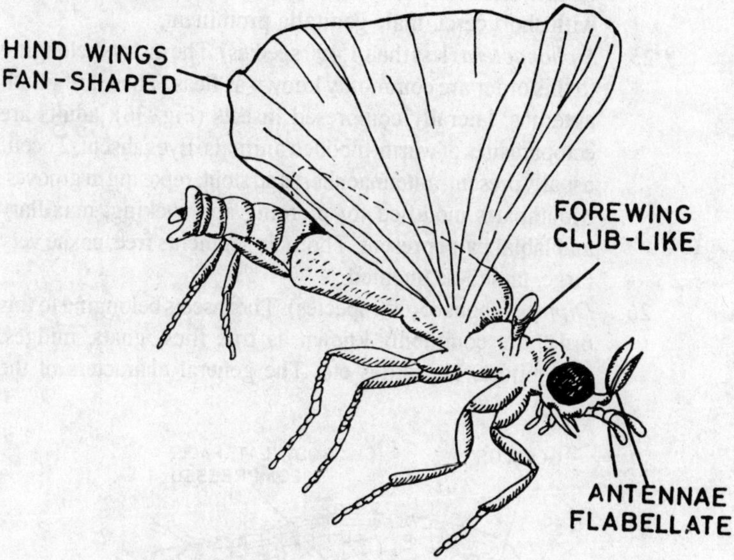

FIG. 34: *Eoxenos* sp.—Strepsiptera (After Pruthi, 1969)

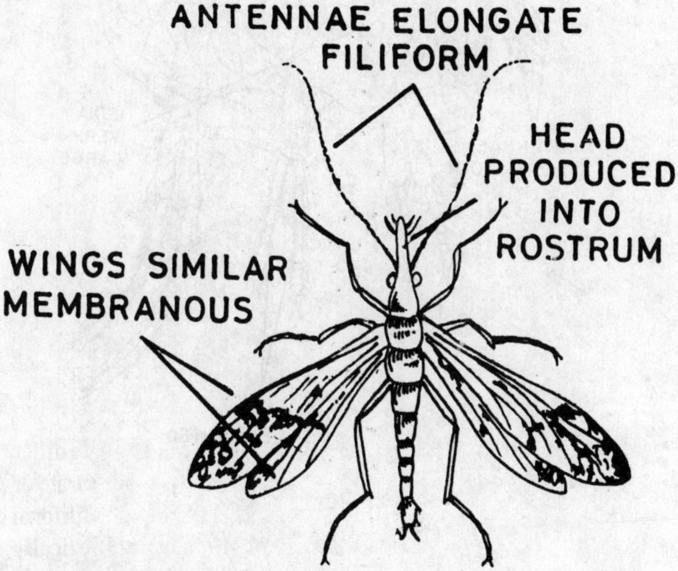

FIG. 35: *Leptopanorpha* sp.—Mecoptera (After Pruthi, 1969)

Rs dichotomously branched, Cu_1 simple, Abdomen elongate with short cerci, male genitalia prominent.

25. *Siphonaptera* (less than 1,400 species). The insects belonging to this order are commonly known as fleas. These are small, apterous, laterally compresed insects (Fig. 36), adults are ectoparasites of warm-blooded animals. Eyes absent, 2 ocelli usually present, antennae short and stout, reposing in grooves. Mouthparts modified for piercing and sucking, maxillary and labial palpi present. Thoracic segments free; coxae very large, tarsi 5-segmented.

26. *Diptera* (about 85,000 species). The insects belonging to this order are commonly known as true flies, gnats, midges, mosquitoes, houseflies etc. The general characters of the

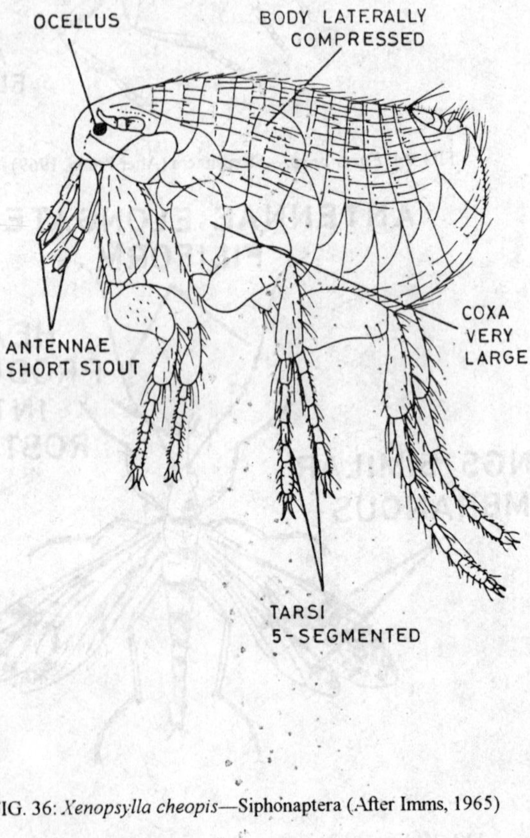

FIG. 36: *Xenopsylla cheopis*—Siphonaptera (After Imms, 1965)

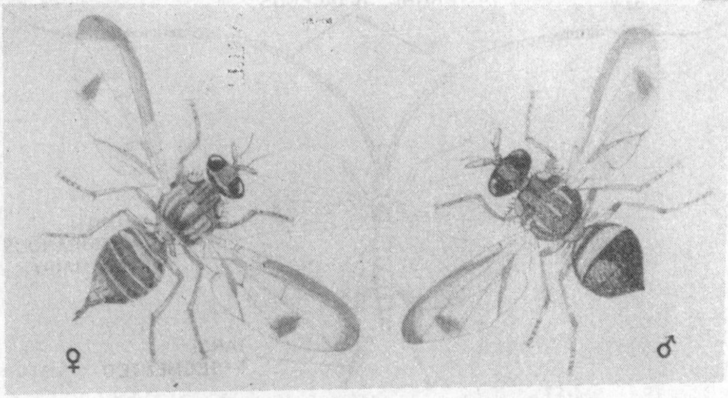

FIG. 37: *Dacus cueurbitae*—Diptera

FIG. 38: *Papilio demoleus*—Lepidoptera

order are : insects with a single pair of membranous wings, the hind pair modified into halters. Mouthparts suctorial, usually forming a proboscis and sometimes adapted for piercing, mandibles rarely present, labium, usually distally expanded into a pair of fleshy lobes (Fig. 37). Prothorax and metathorax small and fused with the large mesothorax. Tarsi commonly 5-segmented.

27. *Lepidoptera* (more than 100,000 species). The insects belonging to this order are known as moths and butterflies

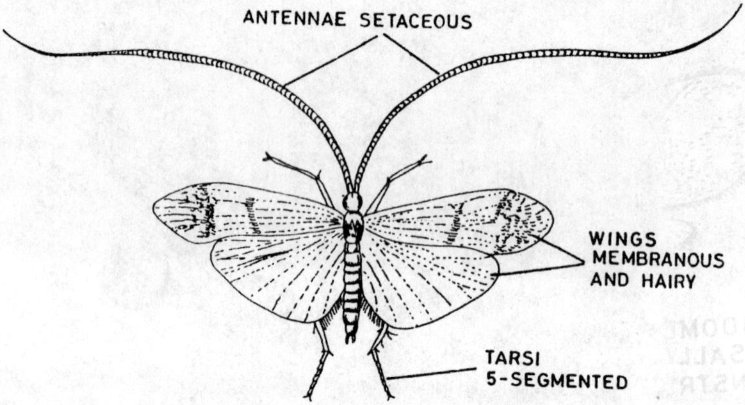

FIG. 39: *Macronema* sp.—Trichoptera (After Lefroy, 1923)

(Fig. 38). These insects are with 2 pairs of membranous wings, cross-veins few in number, The body,wings and appendages clothed with broad scales. Mandibles almost always vestigial or absent, and the principal mouthparts generally represented by a suctorial probascis formed by the maxillae.

28. *Trichoptera* (5,000 species). The insects of this order are known as caddis flies. This order consists of small to moderate-sized moth-like insects with setaceous antennae (Fig.39). Mandibles vestigial or absent. Maxillae signle-lobed with elongate palpi. Labium with a mediuan glossa and well developed palpi. Wings membranous, more or less densely hairy and held roof-like over the back in repose. Fore wings elongate, hind wings broader with a folding anal area. Cross-veins few. Tarsi 5-segmented. Larvae aquatic, more or less eruciform and usually living in cases. Body terminated by hooked caudal appendages.

29. *Hymenoptera* (more than 1,00,000 species). Insects of this order are commonly known as wasps, sawflies, ants, bees, Ichneumon flies etc. These insects have two pairs of membranous wings, often the venation greatly reduced, hind wings smaller than the fore wings and interlocked with the

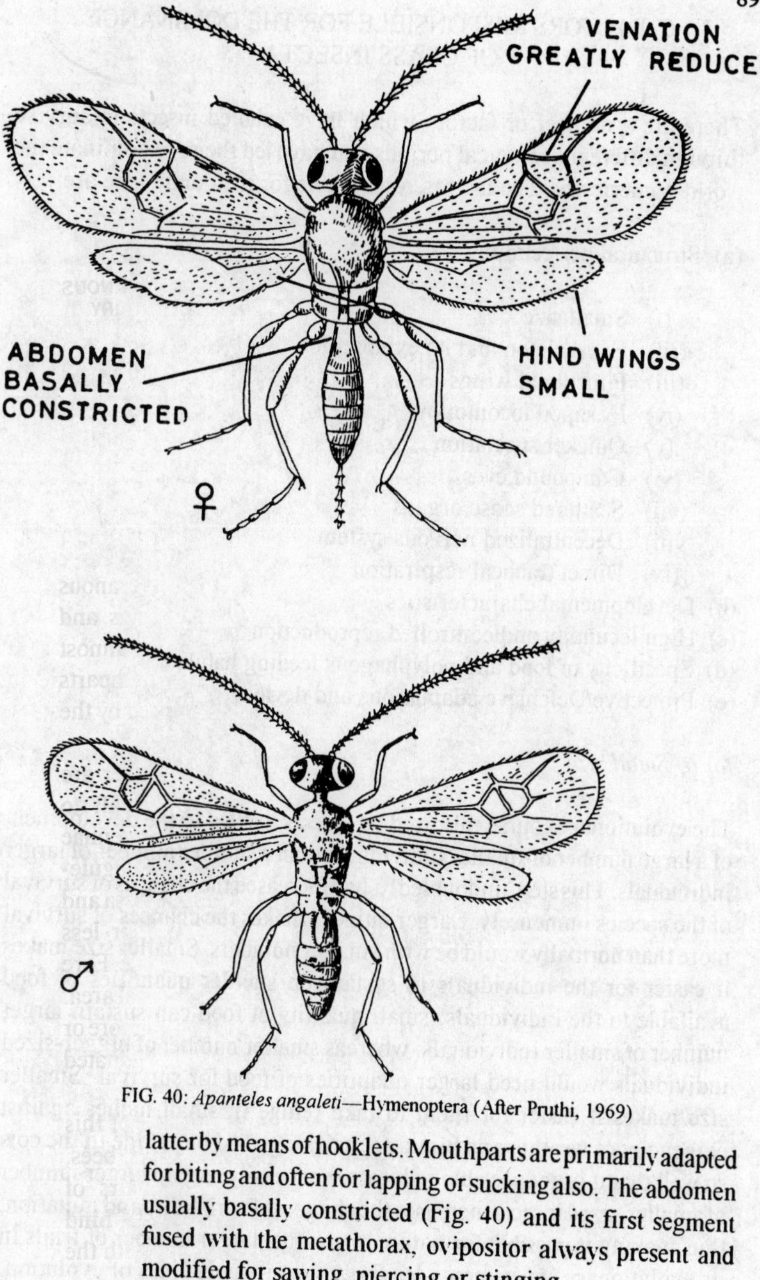

FIG. 40: *Apanteles angaleti*—Hymenoptera (After Pruthi, 1969)

latter by means of hooklets. Mouthparts are primarily adapted for biting and often for lapping or sucking also. The abdomen usually basally constricted (Fig. 40) and its first segment fused with the metathorax, ovipositor always present and modified for sawing, piercing or stinging.

C. FACTORS RESPONSIBLE FOR THE DOMINANCE OF CLASS INSECTA

There are a number of factors which have enabled insects to survive through different geological periods and have led them, to dominate the world numerically. The factors responsible for the dominance are:

(a) Structural perfection

- (i) Small size
- (ii) Exoskeleton and its exploitation
- (iii) Functional wings
- (iv) Hexapod locomotion
- (v) Quicker speciation
- (vi) Compound eyes
- (vii) Scattered sense organs
- (viii) Decentralized nervous system
- (ix) Direct tracheal respiration

(b) Developmental characteristics
(c) High fecundity and controlled reproduction
(d) Specificity of food and polyphagous feeding habits
(e) Protective/Defensive adaptations and devices

(a) (i) Small size

The evolutionary course followed by insects has led to the development of a large number of smaller individuals than a smaller number of larger individuals. This step, undoubtedly, has increased the chances of survival of the species immensely. Larger number makes the chances of survival more than normally would be with smaller numbers. Smaller size makes it easier for the individuals to sustain on smaller quantities of food available to the individuals; small quantity of food can sustain larger number of smaller individuals, whereas smaller number of bigger-sized individuals would need larger quantities of food for survival. Smaller size makes it easier for them to take refuge in small niches against exigencies of weather and biotic agencies e.g., bed bugs hide in the cots so well that it is not possible to locate them easily. Again larger number of smaller-sized insects increases the chances of variability and mutation, thus making it possible for nature to conduct larger number of trials in its evolutionary experiments leading to many more lines of evolution. Smaller size leads to greater overall efficiency in the individual because smaller sized insects have got smaller muscles which are more efficient. It is a scientific fact that power of contraction of muscles is proportional to the square of surface area e.g., it is only two dimensional (L x B) where as the weight of the body is three dimensional (L x B x H) hence increase in the size of muscles increases the weight but not so much the surface area. The implication of this is that the muscles of insects are much more efficient as compared to those of man. A grasshopper leg can exert a power of 20,000 times their own weight and thus outperform human muscle tenfold (Metcalf and Flint, 1979). This is the reason that a flea beetle can jump 33 cm in length and 20 cm in height. If a man were to do it, he would have a long jump of 213 m and a high jump of 137 m which obviously cannot be done hence man cannot compete with the efficiency of flea beetle.

Thus, out of the three different criteria i.e., size, variety and population in which the relative dominance of any particular class can be assessed, the class Insecta can unquestionably claim dominance on the basis of atleast two criteria i.e., variety and population. Apparently they appear to be at a disadvantage so far as their individual size is concerned, but not only do they make up this apparent deficiency by the number and variety but in addition, they actually derive some advantage out of their small size. Due to small size of the insects, the predators do not consider it worthwhile to attack them, for the energy required for hunting them is often more than the amount of energy they are likely to get from it. Insect size is not too small to be invisible. Because of the small size, insects do not create a visual impression of seriousness except in a few cases. Thus, insects exploit their small size in their struggle for existence.

Size of some of the very small-sized insects is given below:

Coleoptera

Family: Ptiliidae

Namosella fungi 0.25 mm

(Imms, 1965)

Hymenoptera

Family : Trichogrammatidae

Trichogramma minutum 0.37 mm

(Matcalf and Flint , 1979)

Mymaridae : *Alaptus magnanimus* 0.21 mm

(Imms, 1965).

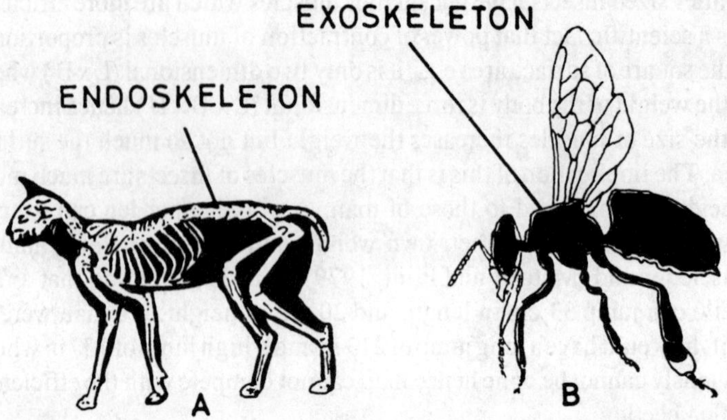

FIG. 41: Endoskeleton of a vertebrate (A) and Exoskeleton of an insect (B)
(After Ayyar, 1940)

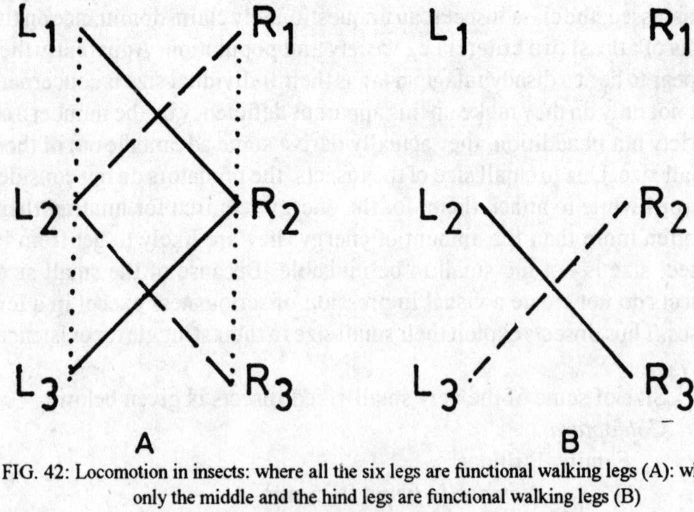

FIG. 42: Locomotion in insects: where all the six legs are functional walking legs (A): where only the middle and the hind legs are functional walking legs (B)
(After Chapman, 1971)

(ii) *Exoskeleton and its exploitation*

This characteristics is shared with other arthropods. Mere possession, however, of exoskeleton and for that matter of any character cannot lead to dominance. It has to be seen what role it plays in the life of the animal and the extent to which it has been able to exploit it. Insects have exploited the mechanical advantage of the exoskeleton to the maximum extent.

(1) The exoskeleton of insects is a strong covering of body. Since it is external in position (Fig. 41B) it provides a much larger surface area for the attachment of muscles than it would have provided if it were internal in position (endoskeleton, Fig. 41A).

(2) Affords protection to the insect in general and to its muscles in particular against mechanical injury. Had it been inside the body it would not have been in a position to afford any protection from mechanical injury atleast to muscles.

(3) *Affords protection against desiccation:* Presence of exoskeleton is an excellent mechanism to prevent desiccation (there are other mechanisms also to protect insects from desiccation). The upper layer of the cuticle is cement layer under which lies the wax layer. Both these layers prevent entry as well as evaporation of water from the insect body. But for this very effective water proofing mechanism of the insect exoskeleton, probably such minute creatures, as majority of insects are , could not have evolved so successfully even as terrestrial forms much less to say as aerial forms. The main disadvantage of small size normally is comparatively much larger surface area per unit body weight and thus greater vulnerability not only to mechanical injury but also to desiccation. Both these problems have been solved successfully in insects on account of the possession of exoskeleton.

(4) *As tool:* Exoskeletal appendages have been variously modified to act as efficient tools for digging e.g., forelegs of Gryllotalpa, (Fig. 20), for catching and holding the prey e.g., forelegs of preying Mantis (Fig. 24) , ovipositor for ovipositing the eggs in plant tissues e.g., fruit flies, mustard sawfly and cotton jassid, in animal tissue e.g., parasitic Hymenoptera, and in soil e.g., locusts and grasshoppers; as aedeagus for mating (slight variation in which leads the pair trying to mate to separate), as mandibles used for biting, cutting and chewing and also as means of defence e.g., all mandibulate insects and mandibulate soldiers of termites.

(5) Maintains shape of the body much more efficiently than endoskeleton. On this account old age is unable to put its stamp on the shape of the body which remains unchanged throughout the life of the insect and even after death. This appears to have helped the insect to a certain extent in its

struggle for existence as it affords protection against predators.

(iii) Functional Wings

Insects are the only invertebrates which possess wings. There are only two other groups of animals possessing wings i.e., birds and bats. Insects do not share this character with other arthropods and, therefore, with regards to this feature insects have a definite superiority over other arthropods.

The power of flight has greatly increased the statistical chances of survival and dispersal except on the wind -swept islands where reverse adaptation in losing the functional wings has taken place. Presence of functional wings has greatly increased the chances: (1) of avoiding inclement weather/climate (2) of feeding and breeding range (3) of eluding natural enemies on the one hand while on the other made it easier for the predatory ones to attack even fast moving prey (4) of speciation.

On account of wings it became easier for insects to move to long distances in short time in search of shelter, food or mate. It is also able to avoid inclement weather/ climatic conditions because being a cold-blooded creature its body temperature also changes with changes of atmospheric temperature. This is why migration takes place. Migration of *Agrotis* on the approach of summer in the plains to hills is a very distinct example of migration due to climatic changes. Had there been no wings most of them would have died and hardly any would have survived even on aestivation in extremely hot places like Delhi and Agra. Another very common example of migration due to climatic changes is that of aphids from plains to hills in March-April. As a matter of fact wings develop greatly in March-April to enable them to migrate to cooler regions for the survival of the species. This has enhanced their chances of survival. Aphids, in their apterous forms are devoured in large numbers by the grubs and adults of the lady bird beetle and the grubs of syrphid fly; because of the absence of wings they are not able to escape the predators. On account of the development of wings they have been able to distribute in practically all ecological pockets. The phenomenon of migration is common in birds also.

Presence of wings has increased the feeding range. If in a particular locality, their food plant is not available or it has been fully eaten up and exhausted they may fly to other places to look for their host plants. This is of great help especially when the food plant is available in small quantities. In contrast to this if they were to depend only on their legs for movement, it can be well imagined how much would they be able to walk in a day coupled with the risk of getting crushed under the feet of men and other animals or would have faced starvation. It is specially helpful during food scarcity and the danger of death due to starvation is avoided e.g., locusts, *Agrotis* etc. The female mosquito which rests on vegetation along with its vegetarian male is able to fly and suck the blood of its hosts only because of the presence of functional wings. Thus increased feeding range has undoubtedly increased the chances of better survival.

Likewise the presence of functional wings has increased their breeding range especially where the breeding medium may have been in small quantities. It has also reduced very much the competition in the use of breeding medium amongst members of the same species. Thus, it is due to the presence of wings only that the female mosquito is able to fly and search the still water for laying eggs as its larvae are aquatic. Similarly housefly and the fruit fly are able to locate their breeding media. Likewise *Trichogramma* is able to search and locate the sugarcane borer eggs for parasiting by layings eggs in the sugarcane borer eggs. Thus, the increase of feeding and breeding range has enabled the survival of even the monophagous species like the borers of sugarcane.

Chances of finding a mate and mating have not only become easier but in some cases they are able to compete as in the case of social insects in nuptial flights. Also, had they no wings, in the case of scarcity of the individuals of either sex, many would have remained unmated and died without leaving their progeny behind and thus reducing the chances of survival of the species.

Presence of functional wings has also given them a new means of eluding the enemy when the life is endangered (insects with the help of their compound eyes can perceive the approach of some object and quickly fly away before being caught by the enemy). If it were to depend only on its legs to move it would have been caught and got killed as it happens with bed bug and other apterous insects. Likewise for the predatory insects it has become easy to follow a fast moving or even flying prey and catch it. Obviously the winged forms have distinct advantage over their apterous counterparts as is evident from the fact that the number of species of winged forms is much larger than that of the apterous ones.

Suitability of Insects for Aerial Existence

(1) Insects have more surface area per unit body weight. Increase in surface area increases evaporation from body surface but because of water proof exoskeleton (epicuticle) the loss of

water due to evaporation is checked.

(2) The spiracles which are another source of water evaporation, are guarded by valves on account of which the opening and closing of spiracles is regulated and, thus, allowing the insects to live on optimum amount of water.

(3) There is also mechanism to conserve water by means of Malpighian tubules with cryptonephridial arrangement for uptake of water from rectum.

Thus, increased surface area could have led to increased evaporation which would have led to desiccation. This has been very effectively checked,as explained above, by the presence of water proof exoskeleton, valves of spiracles, cryptonephredial and water reabsorption mechanisms.

(iv) Hexapod Locomotion

Locomotion means moving from one place to another of one's own accord without the aid of any external agency. Plant kingdom is devoid of this character and whatever movement from one place to another is observed in some plants is not without the aid of external agencies. Locomotion is the characteristic feature of only animal life including insects.

There are different organs which help the animal in locomotion, the most common organ being the leg. All arthropods have paired jointed legs but insects alone have three pairs of jointed legs - an optimum number which no other animal group has hit upon. When an insect walks, the movement of legs takes place in a difinite sequence (Hughes,1952) based on two general principles (1) no leg is raised until the leg behind it is in a supporting position (2) movement of the two legs of a segment alternate (Chapman,1971) i.e., both the legs of a segment do not move together but only one leg of a segment moves at a time. The fore legs serve as organs of traction and pull the body forwards, the middle legs act as support, whereas the hind legs exert propulsive force, pushing it from behind. The movement in terrestrial insects, takes place in a zigzag manner by the formation of two alternating tripods R_1, L_2 and R_3 alternate with the tripod L_1, R_2 and L_3 (Fig.42A). During the course of locomotion both the tripods are on the move and at rest alternately. Thus, the insect never has less than three legs on the ground and can stop at any point without losing its balance, since the three legs enclose the vertical axis through the centre of gravity and stability is enhanced by the fact that the body is slung between the legs so that the centre of gravity is low

(Chapman, 1971). Thus six is evidently the smallest number for the formation of two alternating tripods for stable movement and stable balance , in the same way as small children are able to balance a tricycle whereas for bicycle, they have to learn and practice to balance. Cinematographic analysis of Hughes (1952) also shows that the commonest (but not invariably) order of insect leg movement is R_1, L_2 and R_3 the number of legs on the ground may, at any time be 3, 4 or 5. Increased speed results in increased rate of movement of legs and there may be a change in the sequence but the pattern of behaviour remains fundamentally the same throughout the insect world. (Hughes, 1952). Speed of insect movement varies from insect to insect but in general the rate of movement is higher at higher temperature (Chapman, 1971). Speed also depends upon the size of legs to some extent; with larger legs longer paces can be taken, as for the same frequency of pacing they will move farther than insects with smaller legs. Thus, *Blattella* sp. nymphs move @ 3cm/sec, whereas their adults move @ 20 cm/sec.

In insects having only two pairs of functional walking legs e.g., in *Mantis* (Dictyptera: Mantidae) and *Gryllotalpa* (Orthoptera: Gryllotalpidae) locomotion takes place by the movement of middle leg of one side and hind leg of the other side alternating. Thus, the movement can be expressed as L_2, R_3 alternating with R_2, L_3 (Fig. 42B).

(v) Quicker Speciation

Due to smallness of size, a large number of insects are produced and because of being winged their distribution in different ecological niches becomes easier, mating also becomes easier, variation becomes more frequent with even slight changes; even slight change in the chitinous external genitalia would cause the individual trying to mate to separate and avoid interbreeding and new species are evolved. Thus, smallness of size, large number, presence of functional wings and external genitalia have led to quicker speciation amongst insects.

(vi) Compound Eyes

Insects share the presence of compound eyes only with crustaceans. The compound eye instead of having a single cornea of ordinary eyes, is made up of a number of hexagonal transparent ommatidia or facets each facet

---

Note: R and L stand for right and left legs and 1, 2 and 3 stand for fore, middle and hind legs respectively.

acting as an eye. Thus, the overall effect of a large number of ommatidia is much better than the presence of a single pair of simple eyes. When an object moves, the insect is able to perceive the movement and flies away immediately. Another advantage of the compound eye is that if by chance one, two or a few ommatidia get injured, the insect still continues to have power of perception and does not become totally blind. In addition, the ocelli of some insects probably serve for perception of light and colour. The number of ommatidia present in some insects (Imms, 1965)is as follows:

Housefly *(Musca)*	4,000
Ants	100—1,200
Lepidoptera	12,000—17,000
Odonata	10,000—28,000

(vii) Scattered Sense Organs

Sense organs in insects are scattered on/in different parts of body so that if one part of the body is injured, all the senses of the insect may not cease to function. Thus, dorsal ocelli, lateral ocelli and compound eyes are situated on the head; auditory organ in case of Tettigonidae (long-horned

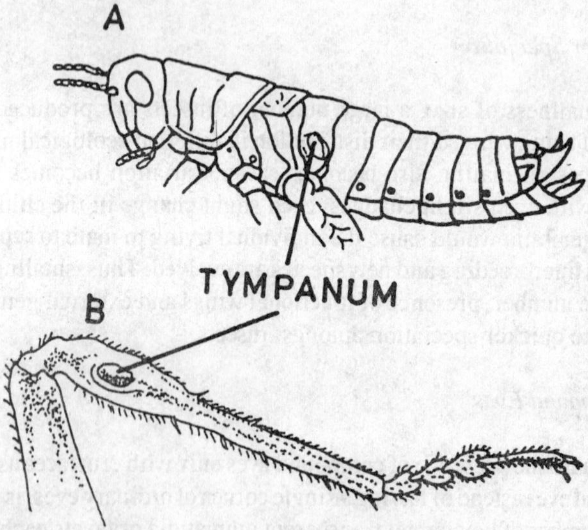

FIG. 43: Tympanum (A) on the first abdominal segment of a locust and (B) on the tibia of a cricket (After Imms, 1965)

grasshoppers) is situated on the first abdominal segment on both sides (Fig. 43A). In Gryllidae it is situated on the tibia of the forelegs (Fig. 43B). In the superfamily Noctuoidea (includes family Noctuidae) tympanal organs are situated on the posterior part of metathorax. In *Phormia* (Diptera) stimulation of tarsi leads to extension of proboscis. Contact chemoreceptors in locusts (Orthoptera) are situated on the ovipositor, which, while depositing eggs in sand, can detect the various inorganic salts and if the concentration of salts be high, it withdraws the abdomen and digs elsewhere. In parasitic insects also chemoreceptors are situated on the ovipositor. Receptors concerned with the perception of irritant substances in grasshoppers appear to be widely distributed over the body especially on the antennae, maxillary palps, legs and elytra. Olfactory receptors are situated on the antennae of *Melanoplus* (Orthoptera) and other grasshoppers (Chapman, 1971) and on the antennae of insects of all major orders. *Eumenis* (Hymenoptera) male has scent patches on the upper side of the forewings and which when touched by the antennae of the female leads to mating. Thus, sense of smell helps in locating the food, oviposition site and mate.

Taste receptors are known to occur on the antennae in some Hymenoptera, on the mouth parts of many insects, on the ovipositor in some, on the tarsi and distal parts of tibia in many Lepidoptera, Diptera and Honeybees (Imms, 1965).

(viii) Decentralized Nervous System

Central nervous system is a ladder-like or string-like chain of ganglia along the ventral side of the body except one complex nerve centre in the head (the brain) which lies dorsal to the alimentary canal. This system (nervous system), despite the highly-evolved stage of insects as a class, is decentralized. Each segment of the body, with its ganglion, is capable of autonomous reflex behaviour that is to say centralization of function is less well-developed as compared to vertebrates. The result of this is that brainless insects can be artificially stimulated to walk, fly or feed .Even when the living abdomen of a female silk moth is separated from the thorax, it can be fertilized by the male and the female thus fertilized can lay eggs (Imms, 1965). In Mantids, copulatory movements of the male and movement of the ovipositor of the female can be made even after head is detached (Imms, 1965). Headless *Schistocerca* can keep a leg flexed so as to avoid electric shocks. How far this absence of centralization of the functions of the nervous system is actually useful in the preservation of the species in nature needs further studies.

(ix) Direct Tracheal Respiration

In higher animals, air enters through nostrils and has to pass through blood, O_2 is absorbed and carried to different parts of the body by blood vessels and the rest of the gases are exhaled. In case of insects, however, this important physiological function is independent of the mediation by blood which in this case, is not required to transmit the gases of respiration. In insects, these gases enter through several paired openings—the spiracles on the surface of the body rather than a single pair of nostrils. These paired entrance openings are connected directly to air-tubes—the trachea and tracheoles, whose network is there all over the body. This mechanism, thus, separates the nutrition and oxygenation system. This system fits in very well with the requirements of terrestrial and aerial existence of minute creatures liable to dessication which could have naturally altered the consistency of liquid blood if the respiration were not direct (Fig. 44 A,B).

This direct tracheal respiratory system, however, puts a limit on the size of the members of the class as there are limitations for air being carried directly into the tracheal network as no pumping mechanism like that of higher animals, exists. This is perhaps the main reason why insects are unable to attain big size and we never come across insects any where near the size of even a cat leave alone horse and elephant.

(b) Developmental Characteristics

Generally in the development of an insect where metamorphosis takes place there are two non-feeding, one or two feeding stages and one reproductive stage. Thus, there are four stages:

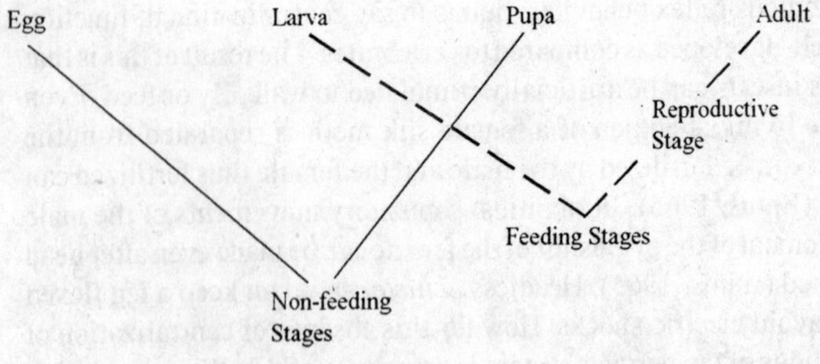

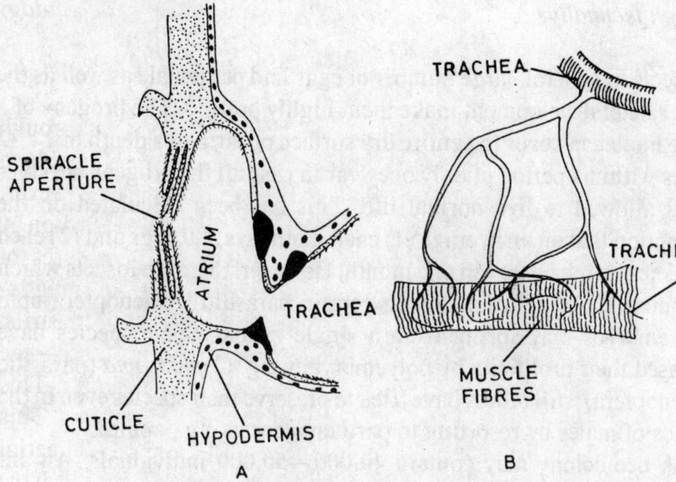

FIG. 44: (A) Spiracle of a lepidopterous larva (After Imms, 1965)
(B) Tracheoles running to a muscle fibre (After Wiggldsworth, 1970)

(i) The two feeding stages i.e., larva and adult are so different in their structural details, food requirements and habitat requirements that there is no internal competition between the parent and offspring regarding food and shelter e.g., lepidopterous larvae are generally leaf eating or plant feeders whereas adults are only nectar feeders. Likewise food of *Athalia* grubs is different from that of their adults.

(ii) The same individual is able to exploit atleast two different kinds of food and habitat in two different stages of its development.

(iii) Interpolation in between two active stages, larva and adult, of an inactive pupal stage with practically no food requirement, the two active stages are best fitted to the conditions of seasons, etc. (optimum for both the stages is different). Thus, the mechanism of metamorphosis of insects has enabled them to adapt themselves to varied situations of existence much better than other organisms, which lack this mechanism.

(c) High Fecundity and Controlled Reproduction

In the Animal kingdom as a whole, as a general rule, the number of eggs laid/young ones born is inversely proportional to the security.

(i) High fecundity

In insects in general, large number of eggs laid per female as well as the quick rate of development make them highly prolific. The progeny of a single moth can cover the entire dry surface of earth to a depth of 24.32 metres within a period of only one year in case all the progeny survives and is allowed to live normal life. This has been calculated on the assumption that on an average (1) each moth lays 200 eggs and (2) each completes its life history in one month. However, there are insects which lay thousands of eggs/female. In certain parasitic Hymenoptera upto 2000 embryos may spring from a single zygote. Some species have increased their prolificity by polyembryony e.g. *Copidosoma* (parasitic Hymenoptera) ;still others have tried to preserve their species even in the absence of males by resorting to parthenogenesis e.g., aphids.

A bee colony may contain 40,000—50,000 individuals. An ant colony may have a population of 50,000 ants. In some soils there may be 2.5 to 150 million insects/hectare. Locusts, may-flies,thrips and aphids fly together in such large numbers that they sometimes resemble cloud or fog (Pruthi, 1969).

A pair of house flies beginning reproduction in April might be progenitors, if all were to live in, of 191,010,000,000,000,000,000 flies by August and would cover the earth 14.29 m deep (Metcalf and Flint, 1979).

Thus, even if man succeeds in killing insect pests in large numbers, their biotic potential is so great that their populations normally do not change from year to year unless the environmental conditions become adverse for them.

(ii) Controlled Reproduction

While on the one hand, insects have taken recourse to prolific breeding due to lack of security in order to ensure the survival of their species, on the other extreme are the social insects which have complete control over their reproduction and have developed the mechanism to keep themselves and their progeny "adequately sheltered". On account of this security, they are sure of preserving their species hence they have rightly eliminated wasteful reproduction. While making their surroundings safe for their habitation they have developed the technique of preparing their habitations themselves suiting for them to live in e.g., comb, different sizes of cells for different castes of young ones etc., for the entire colony, on account of which the environmental resistance has decreased and the potentiality of reproduction has become confined to a few individuals only without any apparent loss to the community. They have developed such a high degree of controlled reproduction (e.g., in wasps, honeybees) that at a time only one female in the colony is fit to reproduce. Others constitute a caste of sterile females. This is a clear cut division of labour between the reproducing females, the queens and the sterile labouring females, the workers. This represents the highest evolution of social order in the entire animal kingdom including the most highly evolved species, man.

In social insects, all males do not mate and out of thousands only one succeeds in mating. This further ensures good progeny.

(d) Specificity of Food and Polyphagous Feeding Habits

Inspite of the fact that there is diversity of food habits amongst insects, there is also, in many cases specificity of food. A vast majority of insects exercise food preference in their food habits, in some cases, this has developed to the extreme. This is a measure to avoid competition of food between different species of insects. The food specificity has gone further ahead. Certain insects feed on plants, only of a particular family e.g., *Athalia lugens proxima* grubs feed on leaves of only family Cruciferae whereas the chrysomelid beetle *Raphidopalpa foveicollis* feeds on the leaves of only Cucurbitaceae. Further, different insects feed on different parts of the same plant. This further reduces the competition of food. Thus some insects feed only on leaves e.g., most of the larvae of the order Lepidoptera (e.g., caterpillars of *Spodoptera litura)* ,*Pyrilla perpusilla, Aleurolobus barodensis, Hieroglyphus nigrorepletus* etc. Some restrict their feeding only to fruits e.g., maggots of *Dacus* spp. and caterpillars of *Earias* spp. Certain species e.g., *Chilo partellus* restricts its feeding to the sorghum and maize stem;*Chilo suppressalis,* the rice stem borer and stem borers of other crops, restrict their feeding only to stem; *Tryporyza nivella,* the sugarcane top borer and*Atherigona varia soccata,* the sorghum shoot-fly restrict their feeding only to the top portion of the plants. Still others like *Emmalocera paratoxalis,* the sugarcane root borer and root borers of other crops restrict their feeding to roots only. Different feeding stages of the same species of an insect may feed on different parts of the same plant e.g., grubs of the red pumpkin beetle feed on the roots and parts of stem lying below the ground level, but the adults feed on the leaves of cucurbits. Cockchafer beetle grubs feed on roots, whereas their adults feed on leaves. The mosquito females are blood suckers while their males are vegetarian.

On the other hand a large number of insects have developed what is known as polyphagous feeding habit on account of which they can live on practically any food in the absence of their preferred food. Thus starving to death is avoided and survival is assured e.g., *Spilosoma obliqua, Spodoptera litura, Schistocerca gregaria, Periplaneta* sp., *Gryllodes* sp., etc.

The development of food habit has developed in yet another direction. It has developed in the direction of taking such food for which there is almost no competition e.g., cellulose eaten by termites and dung by the dung rollers.

(e) Protective/Defensive Adaptations and Devices

Insects like many other animals have developed ways and means to defend themselves against adverse biotic and climatic factors. These adaptations serve double purpose; on the one hand they afford protection to the insects,on the other they give opportunities for the predators to catch hold of their prey easily. However, the major role of these adaptations is defensive. In some cases these devices afford complete protection against adverse biotic and climatic agencies,while in others the devices deter the enemies from catching them e.g., distasteful butterflies, warning colouration of wasps and disagreeable smell of pentatomid bugs. These devices can be classified into four categories:

(i) Morphological adaptations
(ii) Behavioural adaptations
(iii) Construction of protective structures
(iv) Selection of safety niches

(i) Morphological Adaptations

Exoskeleton plays a role in the development of protective /defensive / preying adaptations and devices also. From the exoskeleton are developed spines, bristles, hairs and scales. Hairs and spines etc., in some cases, eject venomous liquid when disturbed so that the agency trying to disturb them may not attempt again. Larvae with poisonous hairs occur amongst Lepidoptera. In some cases, e.g., caterpillars of the moth *Automeris io* and the brown tail moth *Nygmia phaeorrhoea* have barbed hairs. When the barbed hairs break poison is released (Frost, 1942). Some of the common examples of the hairy caterpillars are *Spilosoma obliqua* and *Euproctis lunata*. The contact of the hairs of *Euproctis* larvae with human skin causes severe irritation. These hairs are connected with the underlying hypodermal poison glands. When the hair comes in contact with skin, tip of the hair is broken and the poison liberated. These poisonous hairs are specially dangerous when they happen to lodge in the cornea of the eye. Like 'hairs', scales also are produced from hypodermal cells and are actually modified form of hairs. The common examples are the waxy coating on the larvae and pupae of the sugarcane white fly *Aleurolobus barodensis,* on mealy bugs, and on various coccids.

Strong mandibles of the mandibulate soldiers of termites which they use to protect the colony by seizing or repelling the intruder, are also an example of exploitation of exoskeleton. In some species of termites, heads of soldiers are specially developed and they are able to block gaps in the walls of termitoria, while the workers are undertaking repairs.

Resemblance to surrounding: Certain insects resemble their surroundings; the resemblance in many cases is so complete that unless the insect moves, it is very difficult to recognize it. A few common examples are: Certain grasshoppers which live generally on soil surface e.g., *Chrotogonous* sp. are dull, soil-coloured, resembling soil-colour wehereas other surface grasshoppers like *Atractomorpha* which feed on leaves are bright leafy-green coloured to escape being caught/killed. A British cerambycid *Clytus arietis,* resembles a wasp in external appearances and exhibits even the internal movements suggestive of a wasp (Imms, 1965).In an African cerambycid *Petrognatha gigas,* upper surface resembles dead velvety moss and its irregular antennae look-like dried tendrils or twigs (Imms,1965). The cerambycid *Batocera* and the bug *Halys dentatus* resemble so much the bark of the tree on which they rest that even the trained eyes often miss them. The leaf insect *Phyllium crurifolium* and the stick insect *Carausius morosus* resemble leaf and twig respectively so much that it is difficult to notice them unless they move.

The dead-leaf butterfly, *Kallima* sp. with folded wings, resembles very closely dead leaves fallen on the ground where it rests (Fig.45).

Resemblance of one species to another: Resemblance of one species to another animal species including insects to get protection from enemies is known as mimicry. Mimicry is distinctly different from cryptic colouration. Some insects imitate other species which are unacceptable to the enemies. Some harmless insects have the colour pattern of stinging wasps and of distasteful butterflies and bugs. Such colour patterns are very conspicuous and the enemies are misled (Pruthi,1969).

FIG. 45: *Kallima* sp.—a butterfly resembling dead leaf when at rest

FIG. 46: Mimicry—(A) Ist form of *Papilio polytes* female mimics *Tros hector* and

(B) Female 2nd form of *P. polytes* mimics *T. aristolochiae*
(Courtesy: Head, Division of Entomology, IARI)

The insect, which is copied is known as 'model'and one which copies is known as 'mimic'. Generally the number of mimic is less than that of the model. Certain spiders (Arachnida) resemble ants (Hymenoptera), mix up in their colonies and keep on consuming them, the ants never getting scared of the spiders for they never recognize them as spiders and consider them as members of their own group.

Amongst Hemiptera,there are several instances of resemblance to the same or other orders of class Inscta. The coreid bug *Dulichius inflatus* (brachypterous form) closely resembles the ant *Polyrachis spiniger* and associates itself with it. Another coreid *Alydus calcaratus* is often found in England in association with the ant *Formica rufa* and other ants whose nymphs resemble closely. *Ceria eumenoides,* a saprophagous dipteran (Diptera: Syrphidae) , resembles the wasp *Eumencs* sp. (Hymenoptera: Vespidae) having venomous sting to defend itself. Thus, when the predator comes across *Ceria eumenoides,* it gets confused with *Eumenes* sp.and does not touch it.

Papilio butterflies are best known for illustrating the phenomenon of mimicry. *Papilio dardanus* has 40 kinds of females each of which mimics a distinct model. Two species of *Tros, T. hector* and *T. aristolochiae* are distasteful butterflies. The female of *Papilio polytes* exists in three forms (a) form which resembles the male of the species *(P. plytes)* (b) first form and (c) second form of the females of *P. Polytes* resemble *T. hector* (Fig. 46A) and *T. aristolochiae* (Fig. 46B) respectively so closely in the markings on their wings etc., that the birds take them to be distasteful species of *Tros* and not *P. polytes* females and they are saved.

Warning colourations : The venomous distasteful insects are generally so gaudily-coloured that they can be easily distinguished from other palatable species, Thus when the colour patterns of some species are similar to each other, learning by the predators to avoid one particular colour pattern also leads to avoidance of the other having similar colour pattern even though they may not be distasteful; on this account palatable species gain some advantage by resembling the distasteful ones. By the habit of avoidance of distasteful species wastage is avoided and the distasteful species get an advantage of better survival.

(ii) Behavioural Adaptations

Certain insects when threatened behave in such a way that the predator/ enemy is frightened or dodged.

Stinging Hymenoptera take up such an offensive attitude when threatened, that the predator gets frightened and avoids it; the stinging Hymenoptera, when adopt a threatening attitude, may even attack the enemy in a group and sting collectively.

Termite soldiers of mandibulate type assume threatening attitude with their powerful jaws outstretched.

Papilio larvae when disturbed, put up a forked organ just back of the head and from which is liberated strong disagreeable asmelling liquid and turn their tails upwards.

Some geometrid larvae (Lepidoptera: Geometridae) e.g., *Biston supressaria* keep themselves erect, sway about and look like snakes and thus keep the enemies away (Pruthi, 1969). Larvae of family Notodontidae (Lepidoptera) have a ventral prothoracic gland from which can be ejected irritating fluid. The fluid ejected by the larvae of *Dicranura vimula* has been found to contain formic acid (Imms, 1965)

Death feigning is a kind of sleep and is common amongst vertebrates and invertebrates both; i.e., amongst reptiles, birds, mammals (not amongst fishes), crustaceans and insects. It is a protective device to dodge the enemies especially the predators, and occurs in practically all orders of insects to a certain extent (Frost, 1942) and has been studied in beetles, water scorpions and giant water bugs. While feigning death, insects usually fall to the ground and appear dead and thus deceive their enemies. In death-feint, the insects draw their leg tightly against the body and remain motionless for varying periods.

Duration of death feint is generally a few minutes but in some cases it may be more.

Giant water bug __________________ 8 hours
(Lethocerus americanus)
Giant water bug __________________ 5 hours
(Belostoma sp.)
Plum curculio ____________________ 3.16 to 6.83 minutes
(Conotrachelus nenuphar)

During death feigning, nervous system is thrown into a state of abnormal inhibition by means of various mechanical stimuli, muscles become tense and appendages can be removed without sign of recovery from the feint; however, decapitation or severing the body between pro and mesothorax can break the death-feint in *Curculio* (Frost, 1942).

Grubs of the mustard sawfly, *Athalia lugens proxima* feeding on leaves, when touched, fall down, contract their body and remain motionless on the soil surface for quite some time as if they are dead.

Other Dodging Devices

Young stages of *Papilio* larvae are such a combination of light black, brown and white stripes/patches that they look like bird dropping. Similarly, adult of *Stenoma algidella* (moth) which is white coloured with faint grey markings, when rests upon a leaf, it resembles bird dropping (Frost,1942) and thus escapes its predators.

Leaf beetle *Lina tremulae* and some other Coleoptera emit offensive fluid which renders them unpalatable to the predators; certain carbids secrete a pungent fluid (liquid or gas) from a pair of anal glands. Eisner *et al.* (1963a) have observed *Chalaenius cordicollis* Kirby, a carabid beetle, emitting a strong phenolic odour when handled ." The secretion is discharged from the anal end as a jet of finely discharged spray".

Certain insects emit obnoxious smell when caught e.g., pentatomid bugs like *Nezara viridula,* coreid bugs e.g. *Leptocorisa,* usually produce nauseous odour (foul smell) on being caught and the predator/insect collector gets discouraged. Bed bug also emits a foul smell when caught.

(iii) Construction of Protective Structures

Among many insects, the construction of lodging for protecting themselves against unfavourable situations in life and 'to provide lodging for the off spring is an instinct'. Certain insects remain in the nests and houses constructed by their own ingenuity in a variety of ways. The instinct to construct a protective lodging for the young ones is particularly noticeable among Endopterygota i.e., insects, which, during their development, pass through complete metamorphosis where the helpless pupa has to be protected from natural enemies and extremes of weather.

Termitoria of termites are protective air conditioned dwellings, there being no parallel of this kind anywhere in the animal kingdom. Ants, bees (Fig. 47) wasps are some other examples of insects which construct dwellings for themselves and their offsprings. Certain insects like bag worms and case worms construct shelter for themselves even in their active feeding stages. These construct tubular cases or bags of leaves, twigs or silk. In certain cases, mature larvae construct cocoons around their own bodies before they enter into pupal stage. The cocoon may be made of earth, silk, fibre, hair or other materials depending upon the natural habitat of the insect (Ayyar,1940).

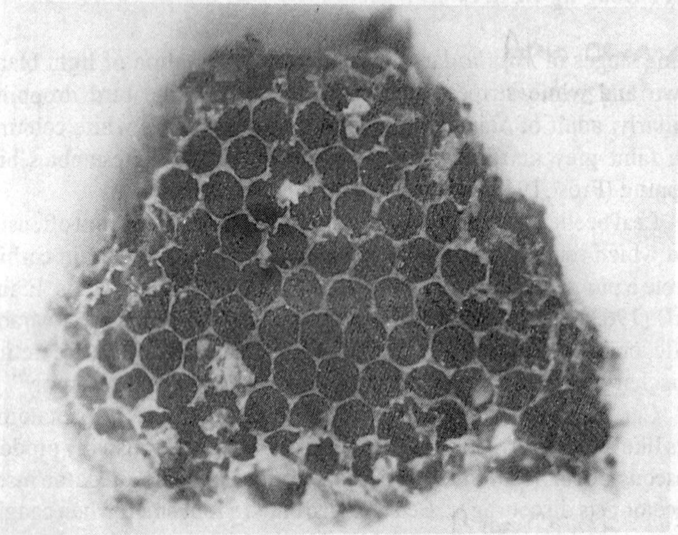

FIG. 47: Comb of honey bee

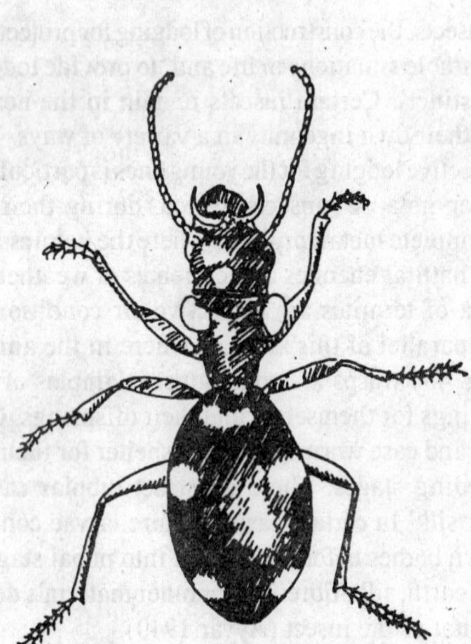

FIG. 48: A carabid beetle (After Pruthi, 1969)

(iv) Selection of Safety Niches

Many insects in the event of danger from predators or on the approach of unfavourable ecological conditions, use their wings to fly away. Some times flying may lead even to large scale migrations e.g., *Coccinella* (beetle), *Papilio demoleus* (butterfly), *Pieris brassicae* (butterfly), *Agrotis* sp. (moth), locusts.

Insects which are neither good fliers nor good at jumping, look for favourable niches e.g., bed bugs, cockroaches, house crickets; bed bugs hide in small crevices of beds so well that they cannot be located easily.

Certain insects have only one, others may have more than one adaptation. Naturally those having more than one adaptation are better protected than those having only one adaptation. A list of some insects having more than one adaptation is given in Table 4.1.

TABLE 4.1. Some insects with more than one protective/defensive adaptations/devices

Sr. No.	Name of Insects	Adaptations/Devices
(1)	(2)	(3)
1.	Locusts and grasshoppers	(i) Colour of these insects simulates the surroundings. Grasshoppers feeding and living on plants and other green vegetation are green coloured. Those living in dry surroundings and sand dunes are grey or brownish coloured with minute spots. (ii) Strong, well-developed, long, hind legs with enlarged femora, adapted for leaping. (iii) Ovipositor adapted for digging holes in the soil/sand and abdomen can expand while ovipositing and contract after oviposition is over. Surface grasshoppers like *Chrotogonus* sp. living on soil surface have flat body with dorsal surface hardly distinguishable from a distance.
2.	*Gryllotalpa spp.*, the mole crickets	(i) Body colour simulates colour of wet soil in which they live. (ii) Anterior pair of legs are fossorial—modified for digging, and very strong with tibiae expanded and digitate.
3.	*Phyllium spp.*, the leaf-insects	(i) Colour and shape both resemble the surroundings very much. (ii) Body is flattened and looks like a leaf. (iii) Legs with leaf-like or stipule-like expansions. (iv) Can remain motionless for long time to evade danger.

TABLE 4.1 *(Contd)*

Sr. No.	Names of Insects	Adaptations/Devices
(1)	(2)	(3)
4.	*Carausius morosus* the walking-stick or the stick-insect	(i) Elongated, cylindrical, rod-like forms, closely resembling twigs-their habitat. (ii) Colour also simulates the surroundings. (iii) Can remain motionless for hours together to evade danger to its life.
5.	*Mantis religiosa* the preying mantis and other mantids.	(i) Colour of the insect simulates that of the habitat. Thus helps the insect partly to protect itself against enemies and mostly to deceive the prey. (ii) Forelegs are modified and adapted for holding the prey, the tarsi fold against tibia and both have strong spines.
6.	*Climex spp.*, the bed bug	(i) Body flat, can easily hide during day in crevices of walls, floor, and furniture and in beds. (ii) Emit foul smell, hence escape being caught.
7.	*Belastoma indica*, the giant water bug	(i) Body flattened, head produced into a beak—adaptations for swimming. (ii) Posterior legs specially adapted for swimming. The tibia are flattened and fringed with hairs.
8.	*Nezara viridula* *Aspongopus janus* Pentatomid bugs	(i) Body colour simulates colour of the surrounding. (ii) They are able to eject obnoxious fluid upto a distance of about 15-30 cm from their metathoracic scent glands and escape beingcaught by their enemies.
9.	*Kallima sp.*, The dead-leaf butterfly	(i) Lower surface of wings is sober-coloured hence the butterfly remains inconspicuouswhile flying. (ii) If while flying it sees danger, it slops flying and rests with wings folded. In the sitting position, it assumes the form of a leaf with a few dark patches, thus resembles fallen-dead-leaves.
10.	*Euproctis lunata*	(i) Entire body of the caterpillar of this insect is densely hairy which do not allow the insecticides to reach their body surface. (ii) 'Hairs', which are modifications of exoskeleton, when come in contact with the skin of other animals, including man, cause irritation. This is protection against its enemies.
11.	Wasps and bees	(i) Generally conspicuously coloured to warn the enemies. (ii) Possess sting through which they inject poison. secreted from their poison glands, in the enemies and escape. (iii) Have the capacity to construct protective structures where they and their progeny live, well protected.

TABLE 4.1 *(Contd)*

Sr. No.	Names of Insects	Adaptations/Devices
(1)	(2)	(3)
12.	*Athalia lugens proxima,* the mustard sawfly	(i) The grubs feed on leaves usually during morning and evening hours and during hot parts of the day remain hiding in soil. (ii) When touched by hand for mechanical picking, they fall down, contract their body and remain motionless for some time and feign death. (iii) The grubs/ pupae aestivate in soil during very hot mo hs of the year to escape unfavourable weather conditions.
13.	Carabids, the predacious ground beetles. (Fig. 48)	(i) Legs are usually long, adapted for running rapidly or are short and thickened suitable for burrowing. (ii) Eject usually volatile liquid from the glands opening above the anus,the discharge is accompanied by distinctly audible sound to frighten the enemy.

5

INSECTS AND HUMAN WELFARE

By and large, insects are considered to be enemies of mankind. A large number of them share man's food by causing damage to underground as well as overground parts of crops right from the seedling to harvesting stage. The process of causing damage does not stop here; they damage stored grains and stored products also. Certain insects act as carriers of viral and bacterial diseases. Still others are household pests eating and polluting edible items in kitchens. Some cause damage to books, hides, furs and cotton, silk, woollen clothes, etc.

A number of insects affect human/animal health by acting as primary/intermediate hosts of human/animal diseases. They may cause injury/disease by serving as mechanical agents in the dissemination of infectious material by biting/stinging and inoculating venom or by harbouring parasites which man may acquire through accidental contact.

In spite of the fact that a large number of insects cause harm to man in a number of ways, man will not be happy if all the insects happen to get killed. There are a number of insects which are beneficial to man in a number of ways so much so that some can be considered to be more or less indispensible to man. Man gets honey, silk and lac from insects. There are insect predators and parasites of harmful insects. There are insect pollinators with whose help man gets better yield. There are certain crops in which fruit setting may not take place at all without insect pollination. Cold light is the gift of insects to man. Man gets certain dyes and medicines from insects and some form a part of human diet also in certain countries.

Table 5.1 gives the names of comparatively more common harmful insects and the way they cause damage. The next table (Table 5.2) gives the names of some more known beneficial insects and their usefulness.

After going through both the tables, you can yourself decide which side of the balance is heavier.

TABLE 5.1 A. Harmful Insects
A. INSECT PESTS OF CROPS
(a) Cereal Crops
(1) Rice *(Oryza sativa)*

Scientific name	Common name	Method of causing damage
Diptera		
Hydrellia philippina Ferino	Rice Whorl maggot	By feeding on unopened central whorl of the leaf.
Orsiolia oryzae (Wood-Mason)	Rice gall midge	By feeding on terminal and auxiilary buds of the young rice plants.
Atherigona exigua Stein	Rice stem fly	By boring the stem of rice seedlings and killing central shoot.
Lepidoptera		
Tryporyza incertulus (Walker)	yellow stem borer	By boring the stem at vegetative stage.
Chilo suppressalis (Walker)	Rice stem borer	By boring the stem at vegetative stage.
C. polychrysus (Meyr.)	Dark-headed rice borer	By boring the stem at vegetative stage.
Sesamia inferens (Walker)	Pink borer	By boring the stem
Chaphalocrocis medinalis (Guenee)	Rice leaf folder	By folding leaves and feeding on green leaf tissues.
Nymphula stagnalis (Zeller)	Rice case worm	By feeding on leaves and cutting off the leaf tips.
Mythimna separata (Walker)	Army worm	By feeding gregariously on the leaves and stem of rice plant.
Hemiptera		
Nilaparvata lugens (Stal.)	Brown plant hopper	By sucking the sap and plugging xylem and phloem with their feeding sheath and pieces of tissue into vessels. It also transmit grassy virus and ragged stunt virus.
Nephotettix virescens (Distant)	Rice green leaf hopper	By sucking the sap and transmitting tungro virus.
Leptocorisa acuta (Thunberg) and	Rice Gundhi bug	By sucking sap from grain at milky stage.

TABLE 5.1 *(Contd.)*

L. *oratoria* (Fabricius)		
Heterococcus rehi (Lindinger)	Rice mealy bug	By sucking plant sap
Coleoptera		
Dicladispa armigera (Olivier)	Rice hispa	By tunnelling and feeding between the two epidermal layers.
Echinocnemus oryzae Marshall	Rice root weevil	By feeding on roots of newly transplanted rice.
Orthoptera		
Hieroglyphus banian Fabricius	Rice grasshoppers	Nymphs and adults both feed on leaves and tender
H. nigrorepletus Bolivar		shoots.
	(2) Wheat (Triticum aestivum L.)	
Diptera		
Atherigona naqvii Steyskal	Shoot fly	By attacking seedlings and killing the central
Lepidoptera		shoot
Sesamia inferens Walker	Stem borer	See under rice
Mythimna unipunctata	Armyworm	See under rice
Haworth		
Coleoptera		
Tanymecus indicus Faust	Gujhia weevil	Grubs feed on roots and adults on growing points.
Myllocerus discolor Boh.	Leaf weevil	Feeding on foliage by adults
Orthoptera		
Chrotogonus trachypterus Blanchard	Surface grasshoppers	By feeding on foliage and tender shoots by nymphs and adults.
Atractomorpha crenulata Fab.		
Thysanoptera		
Anyaphothrips flavicinctus Karny	Thrips	By lacerating tender leaves.
	(3) Maize (Zea mays)	
Diptera		
Atherigona orientalis Schiner	Shoot fly	Maggots bore into growing points of maize seedlings.
Lepidoptera		
Amsacta moorei Butler	Red hairy Caterpillar	By feeding on foliage and other parts of plants.
Euproctis subnotala Walker	Hairy caterpillar	

TABLE 5.1 *(Contd.)*

Chilo partellus Swinhoe	Stem borer	By boring in the stem
C. auricillius Dodgeon	Golden fringe borer (stalk borer)	By boring stalks and internodes.
Sesamia inferens Walker	Pink stem borer	By feeding on leaf whorls and folded central leaves in early stages of development and by boring in the stem.
Mythimna separata (Walker)	Army worm	See under rice
Marasima trapezales (Guen).	Leaf roller	By folding leaves and feeding on leaf epidermis.
Spodoptera litura (Fabricius) and *S. exigua* Hubner	Tobacco caterpillar	By feeding on leaves.
Agrotis ipsilon Rott. *A. spinitera* Hubner	cutworms	Caterpillars cut the seedlings at the ground level.
Helicoverpa (Heliothis) armigera Hb.	Earhead caterpillar	By feeding on maturing grains.
Hemiptera		
Pyrilla perpusilla Walker	Sugarcane leaf hopper	By sucking cell sap from leaves by nymphs and adults.
Rhopalosiphum maidias Fitch.	Maize aphid	By sucking the sap of leaves and shoots and exuding honey dew on which sooty mould grows.
Orthoptera		
Hieroglyphus nigrorepletus Bolivar	Phadka grasshopper	See under rice
(4) Sorghum (Sorghum bicolor)		
Diptera		
Atherigona soccata (Rondani)	Sorghum shoot fly	By cutting the growing point of seedling
Contarinia sorghicola Coquillett	Sorghum midge	By feeding on ovaries.
Hemiptera		
Calocoris angustatus Lethiery	Earhead bug	By sucking sap from tender grain at milky stage.
Rhopalosiphum maidias Fitch	Maize aphid	See under maize
Peregrinus maidis Ashmead	Sorghum shoot bug	By sucking the sap from leaves and whorls.

TABLE 5.1 *(Contd.)*

Lepidoptera		
Chilo partellus (Swinhoe)	Stem borer	See under maize
Sesamina inferens Walker	Pink borer	See under rice
Mythimna separata (Walker)	Army worm	See under rice
Heliothis armigera Hubner	Gram caterpillar	See under peas/Bengal gram.
Agrotis ipsilon Rott. *A. flammatra* Schmffer Mueller	Cut worms	By cutting the plants at ground level.
Amsacta moorei Butler	Red hairy caterpillar	By feeding mainly on leaves.
Eublemma silicula (Swinhoe)	Earhead caterpillar	By feeding on maturing grains.
	(5) Pearl Millet (Pennisetum typhoides)	
Diptera		
Atherigona approximata Malloch	Shoot fly	By cutting growing points and feeding on earheads.
Geromyia penniseti Harris	Pearl millet midge	By eating on ovaries.
Lepidoptera		
Marasmia trapezalis Guenee	Leaf roller	By folding and rolling the leaves near the tip and feeding on the chlorophyll.
Eublemma silicula Swinhoe	Earhead caterpillar	By feeding and webbing the grain.
Coleoptera		
Lytta tenuicollis	Blister beetle	By feeding on pollen and flowers.
Zonabris pustulata Thunberg		
Holotrichia consanguinea Blanchard H. *longipennis* Blanchard H. *serrata* Fabricius	White grubs	Grubs feed on the roots with the result that ultimately the plant dies.
Hemiptera		
Peregrinus maidis (Ashm.)	Shoot bug	By sucking the sap of whorl of central leaves.
Orthoptera		
Colemania sphenarioides Bolivar	Deccan grasshopper	By feeding on leaves

TABLE 5.1 *(Contd.)*

(b) Pulse Crops

(1) Bengal gram Chickpea (Cicer arietinum)

Diptera		
Melanagromyza obtusa Mall.	Pod fly	Maggots bore into the pods and feed on the seeds.
Lepidoptera		
Heliothis armigera Hub.	American bollworm	Caterpillars feed on foliage, flowers and buds and bore into pods and feed on seeds.
Agrotis ipsilon Rott. *A. flammatra* Schiffer-Mueller *A.spinifera* Hubner	Cutworms	See under maize
Laphygma exigua Hubner	Lucerne caterpillar	Caterpillars feed on leaves.
Plusia orichalcea Fabricius	Pea semilooper	do
Coleoptera		
Tanymecus indicus Faust	Gujhia weevil	see under wheat

FIG. 49: *Heliothis armigera* damaging red gram *(Cajanus cajan)* pods

TABLE 5.1 *(Contd.)*

Orthoptera		
Atractomorpha crenulata F.	Surface grasshopper	See under wheat
	(2) Redgram or Pigeon pea (Cajanus cajan)	
Diptera		
Melanagromyza Obtusa	Pod fly	Grubs bore into the tender pods feed on the seeds.
Lepidoptera		
Heliothis armigera Hubner	American boll worm	Caterpillars feed on flowers and buds and bore into pods (Fig. 49)
Exelastis atomosa Walsingham	Plume moth	
Maruca testulalis Geyer	Pod borer	By feeding on flowers and boring the pods. (Saxena 1974.)
Adisura atkinsoni Moore	Pod borer	Same as *M. testulalis*
Cydia critica Meyrick	Leaf binder	By binding the leaves and feeding on these .
Amsacta moorei Butler *A. albistriga* Walker *Euproctis fraterna* Moore	Hairy caterpillars	Caterpillars feed gregariously on the foliage.
E. scintllans (Walker)		
Hemiptera		
Clavigralla gibbosa Spin.	Coreid bug	Nymphs and adults suck sap from leaves, flower buds and developing seeds.
Aphis craccivora Koch *A. cardui* L.	Aphids	Nymphs and adults suck sap from tender shoots flower and young pods.
Coleoptera		
Mylabris pustulata Thunberg	Blister beetle	Adults feed on petals and pollen of flowers.
	(3) Peas (Pisum sativum)	
Diptera		
Phytomyza atricornis Meigen	Pea leaf miner	See under mustard and rapeseed

TABLE 5.1 *(Contd.)*

Melanagromyza		
Phaseoli Coquillett	Pea stem fly	Maggots bore in the outer layer of the stem at the ground level in young plants.
Lepidoptera		
Heliothis sp. Fabricius	Gram caterpillar	By making hole in the pod and feeding on seeds.
Etiella zinckenella Treitsche	Pod borer	-do-
Laphygma exiqua Hubner	Lucerne caterpillar	See under Bengalgram
Hemiptera		
Microsiphum pisi Hubner	Pea aphid	Nymphs and adults suck cell sap of tender shoots, flowers and young pods.

(4) Greengram (Vigra radiata)

Diptera		
Melanagromyza phaseoli Coquillett	Pea stem fly	By mining leaf, petiole and stem by maggots.
Lepidoptera		
Herse convolvule L.	Horn caterpillar	By defoliating the plant
Spilosoma obliqua Walker.	Bihar hairy caterpillar	Feed gregariously and voraciously on foliage.
Azasia rubricans Biosduval	Leaf eating caterpillar	By feeding on leaves.
Coleoptera		
Madurasia obscurella Jacoby	Galerucid beetle	By feeding on foliage.
Pachtychus mungonis M.	Green gram weevil	By eating leaves
Hemiptera		
Empoasca kerri Pruthi	Leaf hopper	By sucking the sap from underside of the leaves.
Aphis craccivora K.	Aphid	See under red gram

(5) Blackgram (Phaseolus mungo)

The insect pests of green gram also infest black gram.

(6) Lentil (Lens culanaris)

There are no specific pests which attack this crop. The important pests of this crop are: *Etiella zinckenella, Agrotis spp., Aphis craccivora.* These have already been given under peas and grams.

TABLE 5.1 *(Contd.)*

	(7) Soybean (Glycine max)	
Diptera *Melanagromyza phaseoli* Coquillett	Stem fly	See under peas
Lepidoptera *Stomopteryx nerteria* Meyrick	Leaf miner	By mining tender leaves and folding the adjacent leaves.
Laphygma exigua Hubner	Lucerne caterpillar	See under Bengal gram
Spodoptera litura Fabricius	Tobacco caterpillar	See under tobacco
Spilosoma obliqua Walker Wilcz	Bihar hairy caterpillar	See under green gram
Coleoptera		
Oberea brevis Gahan	Stem borer beetle	By boring into stem of growing plants by grubs.
Myllocerus spp.	Grey weevils	See under cotton
Hemiptera		
Aphis craccivora Koch	Aphid	See under red gram
Lipaphis erysimi Kaltenbach	Aphid	See under mustard
	(8) Cowpea (Vigna ungniculatar Walp)	
Diptera *Melangromyza phaseoli* Coquillett	Pea stem fly	see under peas
Lepidoptera *Azazia rubicans* Biosduval	Leaf caterpillar	See under green gram
Cosmolyce baeticus L.	Blue butterfly	By feeding on leaves, flowers and pods by caterpillar.
Etiella zinckenella Treitsche	Pea pod borer	See under peas.
Hemiptera *Clavigralla gibbosa* Spinola	Pod bug	By sucking the sap from leaves and developing pod
Aphis craccivora Koch	Bean aphid	See under red gram

TABLE 5.1 *(Contd.)*

	(c) Oilseed Crops	
	(1) Mustard and rape (Brassica juncea and B. campestris)	
Diptera		
Phytomyza atricornis Meigen	Pea leaf miner	By making zigzag mines in the leaves.
Hymenoptera		
Athalia proxima klug	Mustard sawfly	Grubs feed by making holes in leaves
Lepidoptera		
Plutella maculipennis Curtis	Diamond-back moth	By feeding on leaf epidermis. By caterpillar and by making holes
Pieris brassicae Linnaeus	Cabbage butter fly	By feeding on leaves by Caterpillars
Crocidolomia binotoles Zeller	Cabbage leaf webber	See under cruciferous vegetables.
Hellula undalis Fabricius	Cabbage borer	See under cruciferous vegetables
Hemiptera		
Bagrada cruciferarum Kirkaldy	Painted bug	By sucking the sap from plants.
Lipaphis erysimi Kalt.	Mustard aphid	By sucking the sap from leaves, inflorescences, pods,
Myzus persicae Sulzer	Green aphid	and tender stems.
	(2) Groundnut (Arachis hypogaea)	
Lepidoptera		
Amsacta moorei Butler *A. ablistriga* Walker	Red hairy Caterpillars	See under maize
Stomopteryx nerteria Mayrick	Groundnut leaf miner	Caterpillars mine leaves and later joining two leaves and feeding within.
Spilosoma obliqua Walker	Bihar hairy caterpillar	See under green gram
Hemiptera		
Aphis craccivora Koch	Groundnut aphid	By sucking sap from tender shoots and leaves and also by transmitting virus disease.
Coleoptera		
Sphenoptera perotetti Guerin	Stem borer	By boring into the stem near the soil level.

TABLE 5.1 *(Contd.)*

Lachnosterna consanguinea Blanchard	White grub	See under pearl millet
Oxycetonia versicolor Fabricius	Chafer beetle	By eating the root by grubs and leaves, shoots and flowers by adults.
Dermaptera		
Euborellia stoli Dohrn	Earwig	By attacking pods in the soil.
Orthoptera		
Chrotogonus trachypterus Blanchard	Surface grasshopper	See under wheat
	(3) Castor (Ricinus communis)	
Diptera		
Asphondylia ricini Mani	Castor gall fly	By feeding on flowers and tender capsules and pupating inside.
Lepidoptera		
Achaea janata Linnaeus	Castor semi-looper	By feeding voraciously on leaves by caterpillars.
Dichocrosis punctiferalis Guenee	Castor capsule borer	By boring into shoots, pods, and seeds by caterpillars
Euproctis lunata Walker *E. fraterna* Moore	Hairy cater-pillars.	Feed voraciously on leaves
Spilosoma obliqua Walker	Bihar hairy caterpillar	See under green gram
Spodoptera litura Fabricius	Tobacco cater-pillar	See under tobacco
Hemiptera		
Amrasca biguttula biguttula Ishida	Cotton jassid	See under potato/cotton.
Trialeurodes ricini Misra	Castor White fly	By sucking the sap from lower surface of leaves by nymph and adults.
	(4) Safflower (Carthamus tinctorius)	
Diptera		
Acanthiophilus helianthi Rossi	Safflower fly	By feeding on florets and sometimes boring into thalamus by maggots.
Melanogromyza obtusa Malloch	Safflower stem fly	See under Bengalgram

TABLE 5.1 *(Contd.)*

Lepidoptera		
Perigea capensis Guenee	Safflower caterpillar	By feeding on leaves and tender capsule by caterpillars.
Laphygma exiqua Hubner	Lucerne caterpillar	See under Bengalgram
Heliothis armigera Hubner	American bollworm	see under red gram
Hemiptera		
Dactynotus carthami Hille Ris Lambers	Safflower aplid	By sucking sap from leaves by nymphs and adults.
	(5) Sesamum (Sesamum indicum)	
Diptera		
Asphondylia sesami Felt	Sesamum gall midge	By boring into flower buds by maggots and thus producing galls instead of fruits.
Lepidoptera		
Antigastra catalaunalis Duponchal	Leaf and pod caterpillar	By feeding on leaves and boring into shoots and fruits by caterpillars.
Acherontia styx Westwood	Hawk moth	By feeding voraciously on foliage by caterpillars.
Spilosoma obliqua walker	Bihar hairy caterpillar	See under green gram
Amsacta moorei Butler	Hairy caterpillar	See under maize.
Hemiptera		
Aphanus sordidus Fabricius	Pod sucking bugs	By sucking sap from tender pod by nymph and adults.
Eusarcocoris ventralis Westwood		
Deltocephalus dorsalis Motschulsky	Leaf hopper	By sucking juice from plant and transmitting phyllody disease
	(6) Linseed (Linum usitatissimum)	
Diptera		
Dasyneura lini Barnes	Linseed gall midge	By feeding inside the flower buds by maggots.
Lepidoptera		
Laphygma exigua Hubner	Lucerne caterpillar	See under Bengal gram
Agrotis ipsilon Rottemburg	Cutworm	See under sorghum

TABLE 5.1 *(Contd.)*

Heliothis armigera Hubner	American bollworm	See under red gram
Spodoptera litura Fabricius	Tobacco caterpillar	See under tobacco
Euproctis scintillons Walker	Hairy caterpillar	By feeding on leaves and capsules.
Plusia orichalcea Fabricius	Semilooper	See under Bengal gram
Hemiptera		
Myzus persicae Sulzer	Aphid	By sucking the sap from leaves by nymph and adults.

(d) Sugarcane *(Saccharum officinarum)*

Lepidoptera		
Emmalocera depressella Swinhoe	Root borer	By boring in the under ground portion of sugarcane (Fig.50)

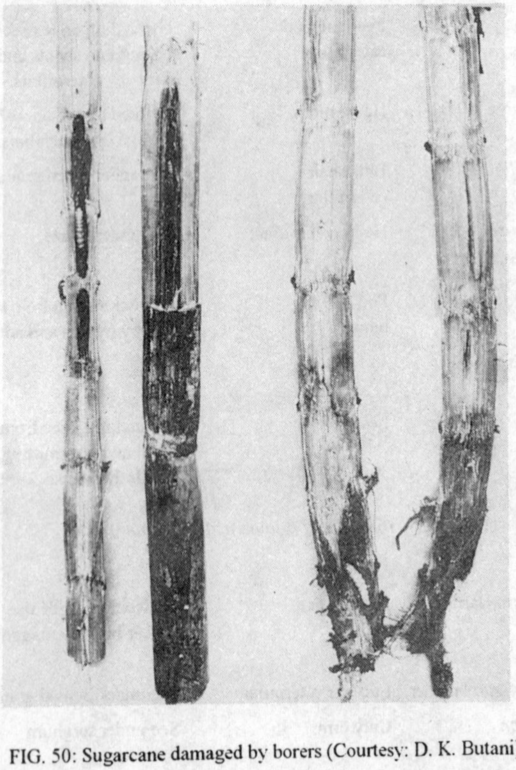

FIG. 50: Sugarcane damaged by borers (Courtesy: D. K. Butani)

TABLE 5.1 *(Contd.)*

Chilo infuscatellus Snellen	Early shoot borer	By boring in the central shoot
C. auricilia Dudgeon	Stalk borer	By mining the leaf sheaths and later penetrating into internodes
C. tumidicostalis (Hampson)	Plassey borer	By boring several holes in the stem.
Tryporyza (Scirpophaga) nivella (Fabr.)	Top borer	By boring central whorl
Sesamia inferens Walker	Pink borer	See under rice

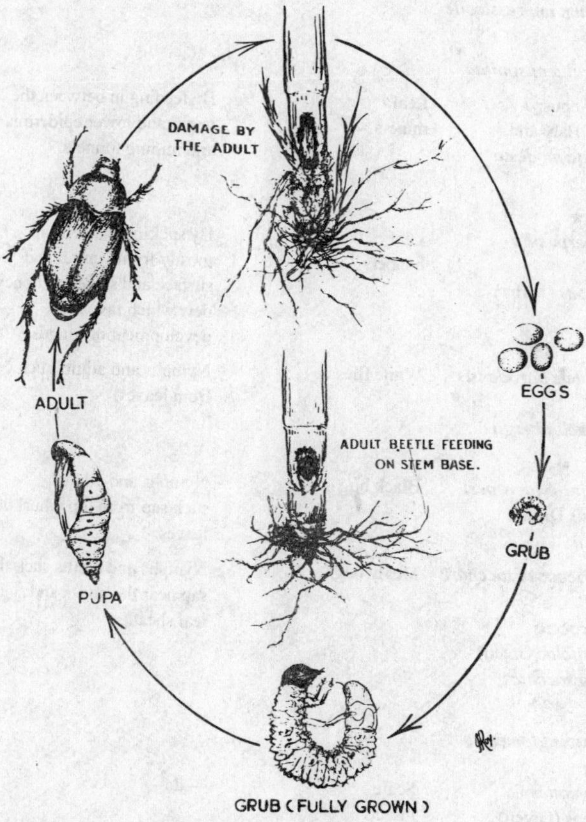

FIG. 51: Sugarcane roots damaged by the white grub, *Holotrichia* sp. and its developmental stages (Courtesy: D. K. Butani)

TABLE 5.1 *(Contd.)*

Ramphimetopus ablutellus Zeller	Green borer	By boring in the shoot
Acigona (=Bissetia) steniellus Hampson	Gurdaspur borer	By feeding gregariously on buds of top internode and boring their way upward.
Mythimna separata (Walker)	Army worm	See under rice
Coleoptera		
Holotrichia consanguinea Blanchard *H. serrata* Fabricius *Alissonotum impressicolle* Arrow	white grubs	Grubs feed on roots (Fig. 51)
Asamangulia cuspidata *Maulik* Leptispa *pygmaes* Baly and *Philodonta modesta* (Weise)	Leaf miners	By feeding in between the upper and lower epidermis and mining tunnels.
Hemiptera		
Pyrilla perpusilla Walker *P. aberrans* (Kirby)	Leaf hoppers	By sucking the cell sap mostly from ventral leaf surface and secreting honey dew which favours development of fungus.
Aleurolobus barodensis Maskell *Neomaskelia bergii* Signorat	White flies	Nymphs and adults suck cell sap from leaves.
Covelerius (Macropes) excavatus Distant	Black bug	Nymphs and adults suck sap from top whorl of the leaves.
Saccharicoccus sacchari (Ckll) *Pseudococcus saccharifolii* (Green) *Ripersia sacchari* Green	Mealy bugs	Nymphs and adults suck the sap near the nodes and leaf sheaths.
Melanaspis glomerata (Green) *Duplachionaspis divergens* (Green)	Scale insects	—do—

TABLE 5.1 *(Contd.)*

Odonaspis sacchariculis (Zehnter)		
Orthoptera		
Hieroglyphus banian Fab. *H. nigrorepletus* Bolivar	Grasshoppers	See under rice
	(e) Fibre crops	
	(1) Cotton (Gossypium spp.)	
Lepidoptera		
Earias insulana Boisduval *E. vittela* Fabricius	Spotted bollworms	By boring into the top shoot and later into bolls.
Pectinophora gossypiella Saunders	Pink bollworm	By boring into bolls and seeds (Fig. 52)
Heliothis armigera Hubner	American bollworm	See under red gram
Sylepta derogata Fabricius	Cotton leaf roller	By rolling the leaves and feeding within
Coleoptera		
Myllocerus maculosus Desbrochers	Grey weevil	By feeding on roots by grubs and margin of leaves by adults
Pempherulus affinis Feust	cotton stem weevil	By boring the stem near the base and causing gall-like swelling.

FIG. 52: Cotton bolls damaged by the pink boll worm, *Pectinophora gossypiella*

TABLE 5.1 *(Contd.)*

Sphenoptera gossypii Cotes	Cotton stem borer	By boring into stem
Hemiptera		
Dysdercus koenigii Fabricius	Red cotton bug	By sucking sap by both nymphs and adults, of flowers, buds and bolls and tainting lint with faeces.
Oxydcarenus lactus Kirby	Dusky cotton bug	By feeding gregariously by both nymphs and adults on sap of immature seeds.
Amrasca biguttula biguttula Ishida	Cotton jassid	Nymphs and adults sucksap from underside of the leaves
Bemisia tabaci Gaennadius	Cotton whitefly	By sucking sap by both nymphs and adults from leaves, and acting as vector of virus disease
Aphis gossypii Glover	Cotton aphid	By sucking sap by nymphs and adults from underside the leaves.
Orthoptera		
Chrotogonus trachypterus Blanchard *C. oxypterus* Blanchard	Surface grass hoppers	See under wheat
Thysanoptera		
Thrips tabaci Lindemann *Scirtothrips dorsalis* Hood	Thrips	By sucking the sap and scraping the green portion of the leaves.

(2) Jute (Corchorus capsularis)

Lepidoptera		
Anomis sabulifera Guenne	Jute semilooper	By feeding on the leaves of the top shoots and apical buds.
Spilosoma obliqua Walker	Bihar hairy caterpillar	See under green gram
Laphygma exigua Hubner	Lucerne caterpillar	See under Bengal gram
Coleoptera		
Apion corchori Marshall	Jute stem weevil	By boring into stem, top shoot or near the axil of the leaves by grubs.

TABLE 5.1 *(Contd.)*

Nodostoma bengalensis Duvivier	Leaf beetle	By feeding gregariously on leaves by grubs.
	(f) Tobacco (Nicotiana tabacum)	
Lepidoptera		
Spodoptera litura Fabricius	Tobacco caterpillar	By feeding voraciously and scraping the leaves by caterpillars.
Gnorimoschema heliopa Lower	Stem borer	By mining through leaves, midribs and finally boring into stem.
Laphygma exigua Hubner	Lucerne caterpillar	See under Bengal gram
Plusia signata Fabricius	Semilooper	Seed under Bengal gram
Heliothis armigera Hubner	American ball worm	See under red gram
Amsacta spp.	Hairy caterpillar	See under sorghum
Agrotis ipsilon Rott.	Cutworm	See under sorghum
Hemiptera		
Bemisia tabaci Gennadius	Whitefly	See under cotton
Myzus persicae Sulzer	Peach green aphid	By sucking the cell sap from underside of the leaves by nymph and adults.
Aphis gossypii Glover	Cotton aphid	See under cotton
Coleoptera		
Mesomorphus villiger Blanchard	Ground beetle	By nibbling the leaves of young plants by adults.
Orthoptera		
Atractomorpha crenulata Fabricius *Acrida exaltata* Walker *Chrotogonus robertsoni*	Surface grasshoppers	By feeding on foliage of tender shoots of the plants by nymphs and adults.
Thysanoptera		
Thrips tabaci Lindemann	Onion thrips	By lacerating leaf tissue and feeding; the leaf tip fades and the plants may ultimately fade and dry (Ananthakrishnan,1969)

TABLE 5.1 *(Contd.)*

	(g) Plantation Crops	
	(1) Coconut (Cocos nucifera)	
Lepidoptera		
Nephantis serinopa Meyrick.	Black-headed caterpillar	By feeding on the under surface of the fronds in galleries of silky thread spun by caterpillars.
Macroplectra nararia Moore	Slug caterpillar	By scraping green matter from the underside of leaves by caterpillars.
Gangara thyrsis Moore	Leaf roller	By rolling the leaves and feeding inside by caterpillars.
Hemiptera		
Stapanitis typicus Distant	Coconut bug	By sucking the sap and by transmitting wilt disease of root.
Aspidiotus destructor Signoret	Scale insect	Nymphs and adults suck sap from underside of the leaves
Coleoptera		
Oryctes rhinoceros Linnaeus	Rhinoceros beetle	By boring into unopened fronds by adults and reaching the growing points and damaging leaves and flower bunches (Fig.53)
Rhynchophorus ferrugineus Fabricius	Red palm weevil	By boring the trunk by grubs and feeding inside.
Leucopholis coneophora Burmeister	Cockchafer beetle	By feeding on roots by grubs
	(2) Tea (Camellia sinensis)	
Lepidoptera		
Biston suppressaria Guenee	Tea looper caterpillar	By feeding on leaves by caterpillars.
Andraca bipunctata Walker	Bunch caterpillar	By feeding voraciously on the foliage by caterpillars
Eterusia magnifica Butler	Tea slug	By defoliating plants by caterpillars(Banerjee, 1983).
Clania crameri Westwood	Bagworm	By making bag-like structure of vegetable matter and feeding on the leaves.
Hemiptera		
Helopeltis theivora	Tea mosquito	By sucking the sap from the

FIG. 53: Coconut tree top damaged by the Rhinoceros beetle, *Oryctes rhinocerus*

TABLE 5.1 *(Contd.)*

Waterhouse		buds and young leaves by nymph and adults.
Taxoptera aurantii Boyr	Tea aphid	Generally suck sap from young shoots (Banerjee, 1983)
Aspidiotus camellae Signoret *Saissetia coffeae* Walker	Scale insects	By sucking the sap of the plant by nymph and adult.
Thysanoptera		
Taeniothrips setiventris Bagnall *Dendrothrips bispinosus* Bagnall	Tea thrips	By lacerating the leaf tissue and imbibing sap.
	(3) Coffee (Coffea arabica)	
Diptera		
Melanagromyza coffeae (Koningsberger)	coffee leaf miner	By mining leaves
Lepidoptera		
Zeuzera coffeae Nietner	Red Borer	By boring into stem by caterpillars.
Agrotis segatum Schiffer Mueller	Cutworm	By feeding on foliage by caterpillars
Hemiptera		
Saissetia coffea Walker	Brown bug	Nymphs and adults suck sap from twigs, leaves and berries and secrete honey dew on which sooty moulds grow (D'Souza *et al.*, 1983)
Coleoptera		
Xylosandrus compactus Eichoff	Shot hole borer	By tunnelling bark by grubs and adults
Xylotrechus quadripes Chevrolat	White stem borer	By boring the main stem and thick branches by grubs
Phyllophaga conferta Sharp	Cockchafer beetle	By feeding on roots by grubs
Araecerus fasciculatus de Geer	Coffee bean weevil	By infesting berries in the field and beans in storage by beetle
Thysanoptera		
Scirtothrips bispinosus Bagnall	Coffee thrips	By lacerating and feeding on leaves (D'souza *et al.*, 1983)
Heliothrips haemorrhoidalis Bouche		

TABLE 5.1 *(Contd.)*

	(4) Rubber (Hevea brasiliensis)	
Coleoptera		
Holotrichia bidentata Bum.	Cockchafer grubs	Grubs feed on tap roots and laterals of young rubber plants in the nursery. They also nibble the finer roots and plants droop and dry up. Adults are not reported to cause damage to rubber plants (Pillai and Jayarathnam,1983)
Minthea rugicollis Wlk. *Heterobostrychus aequalis* Waterb. *Sinoxylon anale* Lense	Boring beetles	Direct attack on healthy plants not reported, attack plants which have suffered initial injury due to other causes, bore into wood making short holes (Pillai and Jayarathnam, 1983).
Lepidoptera.		
Simplicia simulata Mre.	Mulch-eating caterpillar	The caterpillars infest mulch kept to protect nursery plants from scarching sun. No damage to rubber plants reported (Pillai and Jayarathnam, 1983).
Aetherastis circulata Meyr.	Gallery making caterpillar	Make galleries on rubber trunks and branches, generally feed on dead bark, but occasionally on living bark also (Pillai and Jayarathnam, 1983).
Hemiptera		
Saissetia nigra Nietri	Scale insect	Sporadic pest, females suck sap usually from terminal green portions, infestation also reported on mid ribs and petiolesof leaves (Pillai and Jayarathnam, 1983).
Ferrisiana virgata Ckll	Mealy bug	Damage by nymphs and adults reported to young plants in nurseries, Nature of damage similar to that of scale insects.
	(h) Fruit crops	
	(1) Mango (Mangifera indica)	
Diptera		
Dacus dorsalis Hendel	Fruit flies	Females lay eggs under

TABLE 5.1 *(Contd.)*

D. zonatus Saunders *D. correctus* Bezzi		the epidermis of fruits and hatching maggots feed on pulp.
Hemiptera		
Idioscopus clypealis Lethierry *I. niveosparsus* Lethierry, and *Amritodus atkinsoni* Lethierry	Mango hoppers	Adults and nymphs suck cell sap from new shoots, buds and flowers.
Drosicha mangiferae Green	Mango mealy bug	Adults and nymphs suck sap from young shoots and tender leaves (Fig. 54)
Aspidiotus destructor Signnoret *Ceroplastes pseudoceriferus* Green *Parlatoria pergandii* Comstock *Pseudococcus sp.* *Phenococcus mangiferae* Green	Scale insects	Suck sap from leaves and tender shoots (Butani and Srivastava, 1983)

FIG. 54: Mango mealy bug. *Drosicha mangiferae* on a mango twig

TABLE 5.1 *(Contd.)*

Lepidoptera		
Indarbela quadrinotata Walker *I. tetraonis* Moore	Bark-eating caterpillars	Caterpillars feed on bark and produce ribbon-like loose mass and later bore into the trunk making zigzag galleries
Coleoptera		
Batocera rubus Linnaeus *B. rufomaculata* (de Geer)	Mango bark borers	Grubs bore into the bark, stem and shoots
Sternochetus magniferae Fabricius *S. frigidus* Fabricius	Mango stone weevils	By laying eggs in epicarp and boring the stone by grubs and emergence of adults through pulp.
Deporaus marginatus (Pascal)	leaf weevil	Weevils feed on foliage
Apoderus trans-quebaricus Fabricius	leaf weevil	Weevils cut and roll leaves (Butani and Srivastava, 1983)
Hymenoptera		
Oecophylla smaragdina Fabricius	Red ant	Builds nest amongst tuft of leaves, which are webbed together by gummy exudation of ants. Does not feed upon leaves / fruits but causes considerable anoyance to the persons picking fruits (Pruthi, 1969).
	(2) Citrus (Citrus spp.)	
Diptera		
Dacus spp.	Fruitflies	See under mango
Lepidoptera		
Papilio demoleus Linnaeus *P. polytes* Linnaeus	Lemon butterflies	Caterpillars feed on leaves (Fig. 55)
Phyllocnistis citrella Stainton	Citrus leaf miner	By feeding inside leaf tissue and making zigzag galleries
Otheris fullonia Linnaeus *O. materna* Linnaeus *Calpe emarginata* Fabricius	Fruit sucking moths	Moths suck the fruit juice after dusk.
Indarbela quadrinotata Walker	Bark eating caterpillar	See under mango

FIG. 55: Lemon butterfly, *Papilio demoleus* caterpillars feeding on citrus leaves

TABLE 5.1 *(Contd.)*

Chloridolum alcamena Thomson	Stem and shoot borers	By boring into shoot and stem
C. cincium Guerin		
Hemiptera		
Dialeurodes citri Riley and Howard		Nymphs and adults suck
D. citrifolii (Morgan)	Citrus white	sap from leaves and secrete
D. elongata Dozier	flies	a sticky substance on which sooty moulds develop
Aleurocanthus husainii Corbett		

TABLE 5.1 *(Contd.)*

A. citriperdus Quaintance and Baker	Citrus White flies	
A. woglumi Ashby		
A. punjabensis Corbett		
Aleurolobus citrisothi		
Corbett		
A. marlatti (Quintance)		
Diaphorina citri Kuwayama	Citrus psylla	Nymphs and adults suck sap from tender parts of the host (Pruthi, 1969).
Aonidiella aurantii (Maskell) *A. citrus citrina* Coquillett *A. orientalis* (Newstead) *Parlatoria oleae* Clover *p. pergandii* Comstock *p. zizyphis* (Lucas)	Scale insects	Suck the sap from leaves and attack fruits as well (Butani and Srivastava, 1983).
Hymenoptera		
Oecophylla smaragdina Fabricius	Red ant	See under Mango
	(3) Guava (Psidium guajava)	
Diptera		
Dacus dorsalis Hendel	Fruit flies	See under mango.
D. zonatus Saunders		
Lepidoptera		
Indarbela tetraonis Moore *I. quadrinotata* Walker	Bark- eating caterpillars	See under mango
Dichocrosis punctiferalis Guenee	Castor capsule borer	See under castor
Viriachola isocrates Fabricus	Anar butterfly	See under pomegranate
Hemiptera		
Drosicha mangifera Green *Hemiaspid oproctus cinerus* Green *Nipalcoccus viridis*	Mealy bugs	Suck sap from young shoots and tender leaves.
(Newstead)	Scale insects	
Pulvinaria psidii Maskell		

TABLE 5.1 *(Contd.)*

Coleoptera		
Microclona leucosticta Meyrick *M.technographa* Meyrick *Aeolesthes holoserica* Fabricius	Stem borers	Grubs bore into stems and shoots
	(4) Grapevine (Vitis vinifera)	
Lepidoptera		
Sylepta lunalis Guenee	Leaf roller	Caterpillars roll leaves and feed inside
Hemiptera		
Arboridia viniferata *Aspidiotus destructor* Signoret	Jassid Scale insect	Adults and nymphs suck sap from leaves and tender shoots.
Thysanoptera		
Rhipiphorothrips cruentatus Hood	Grape-vine thrips	Adults and nymphs lacerate leaves; occasionally attack fruits also.
Coleoptera		
Scelodonta strigicollis Motschulsky	Flea beetle	A specific pest of grape vine; adults most destructive pest of leaf buds.
Sthenias grisator Fabricius	Girdler beetle	Adults ring or girdle the stem, often killing it (Butani and Srivastava, 1983)
Adoretus lasiophygus Burmeister *A. horticola* Arrow *A. duvauceli* Blanchard *A. versutus* Harold *Anomala dimidiata* Hope	Cockchafer beetles	Adults feed voraciously on leaves, grubs feed on roots.
	(5) Zizyphus (Zizyphus sp. mauritiana)	
Diptera		
Carpomyia vasuviana Costa	Ber fruit fly	Most destructive pest. Female lays eggs under the epidermis of fruits and hatching maggots feed on the pulp of fruits.
Lepidoptera		
Meridarchis scyrodes Meyr.	Ber fruit borer	Caterpillars cause damage to fruits
Thiacides postica		

TABLE 5.1 *(Contd.)*

Walker		
Euproctis fraterna Moore	Hairy caterpillars	Caterpillars feed on leaves.
E. feava Fabricius		
E. lunata Walker		
Beara dichromella Walker		
Inderbela spp.	Bark eating caterpillar	See under mango
Hemiptera		
Perissopneumon tamarindus (Green)	Ber mealy bug	Adults and nymphs suck sap from twigs, branches and leaves.
Coleoptera		
Adoretus pallens Arrow	Cockchafter beetles	Grubs feed on roots, and adults on leaves
A. nitidus Arrow		
A. versutus Harold		
Xanthochelus supercilosis Gryllanhal		
Myllocerus subulosus Marshall	Ber weevils	Adults feed on leaves (Butani and Srivastava, 1983)
M. discolor Boheman		
M. undecimpustulatus Faust		
M. transmarimus Herbst		
	(6) Pomegranate (Punica granatum)	
Diptera		
Dacus dorsalis (Hendel)	Fruit flies	See under mango
D. zonatus (Saunders)		
Lepidoptera		
Virachola isocrates Fabricius	Anar butterfly	By boring the young fruits and feeding on developing seeds and tissues by caterpillars.
Dichocrosis punctiferalis (Guenee)	Castor Capsule borer	See under castor
Othreis fullonia (Clerck)	Fruit sucking moth	See under citrus
Indarbela tetraonis Moore	Bark eating caterpillar	See under mango
Deudorix epijarbus Moore	Pomegranate borer	Caterpillars damage fruits by boring
Euproctis fraterna Moore	Hairy caterpillar	See under maize

TABLE 5.1 *(Contd.)*

Hemiptera		
Siphoninus finitimus Silvestri	Anar whitefly	Nymphs and adults suck the sap from leaves.
Coleoptera		
Myllocerus laetvirens Marshall	Grey weevil	Adults feed on leaf margin and laimina
	(7) Emblic (Emblica officinalis)	
Lepidoptera		
Virachola isocrates Fabricius	Anar butterfly	See under pomegranate
Indarbela tetraonis (Moore)	Bark eating caterpillar	See under mango
Batonsa stylophora Swinhoe	Gall forming caterpillar	Caterpillars form galls on tender shoots (Butani and Srivastava, 1983)
Hemiptera		
Cecrciaphis emblica	Amla aphid	Adults and nymphs suck sap from tender parts of trees.
Nipaecoccus vestator Maskell	Mealy bug	
	(8) Litchi (Litchi chinensis)	
Lepidoptera		
Indarbela tetraonis Moore	Bark-eating caterpillars	See under mango
I. quadrinotata Walker		
Cacoecia epicryta Meyrick	Peach leaf roller	See under peach
Platypapla aprobala Meyrick	Leaf roller	Rolls younger leaves and affects yield (Butani and Srivastava, 1983).
Virachola isocrates Fabricius	Anar butterfly	See under pomegranate
Argyroplace illepida Meyrick	Fruit borers	Caterpillars bore into the developing fruits and damage even the stone.
Cryptophleba carpophaga Walsingham		
C. illepida (Butler)		
Hemiptera		
Chrysocris stollii (Wolff.)	Pentatomid bug	Sucks juice from developing fruits.

TABLE 5.1 *(Contd.)*

Drosicha mangiferae Green	Mango mealy bug	See under mango
Dialeurodes elongata Dozier *Aleurocanthus husanini* Corbett	White flies	Nymphs and adults suck the sap from leaves and tender shoots.
Coleoptera		
Amblyrrhinus poricollis Beheman	Plum weevil	Adults feed on epidermis of leaves
Myllocerus discolor Beheman *M. undecimpustulatus* Faust	Weevils	Adults feed on leaves.
(9) Apple (Pyrus malus)		
Lepidoptera		
Lymantria obfuscata Walker *L. ampla* Walker	Gypsy moths	Caterpillars feed on leaves in years of outbreak,can cause complete defoliation in orchards (Batra, 1983)
Malacosoma indica Walker	Tent caterpillar	Caterpillars feed voraciously on leaves and even the tender bark (Batra,1983)
Cracillaria zachrysa Meyrick	Apple leaf miner and Roller	Caterpillars do tunnelling in the leaves and later roll the margins of leaves (Batra, 1983)
Cacoecia sarcostega Meyrick	Apple leaf roller	Caterpillars feed on leaves, buds and flowers after rolling or webbing them.
Euproctis signata Blanchard	Gold tail hairy caterpillar	Caterpillars feed voraciously on leaves, buds, and flowers.
Carpocaspa pomonella Linnaeus	Codling moth	Caterpillars, on hatching bore into the young fruits through the calyx before it closes and feed on soft pulp (Batra,1983).
Zeuzera sp.	Leopard moth	Caterpillars bore into tender shoots and branches (Batra,1983).
Hemiptera		
Quadraspidiotus perniciosus (Comstock)	San Jose scale	Most serious pest of apple; sucks sap from bark of stem, branches, leaf stalk, fruits stalk and soft twig, surface of leaf and fruit (Batra, 1983).

TABLE 5.1 *(Contd.)*

Pseudaulacapsi pentagona	Scale insect	Besides stem, they also damage branches, foliage and fruit (Batra, 1983).
Eriosoma lanigerum (Hausmann)	Woolly aphis	Nymphs and adults suck sap from stem, branch, shoot, petiole of leaf, fruit stalk,calyx end of fruit and root and covering by cottony wool (Batra, 1983)
Coleoptera		
Dorysthenes hugelli Redtenbacher	Apple root borer	Newly hatched grubs feed on decaying organic matter, then feed on finer roots & later bore into bigger roots. Bigger-sized grubs bore into underground part of stem also.
Apriona cinerea Chever *Batocera rufomaculata* Dejean *Aeolesthes holosericea* Fabricius	Wood borers	Grubs bore into the trunk and branches.
Adoretus duvauceli Blanchard *A.versutus* Harold *Anomala lineatopennis* Blanchard *A. rufiventris* Redtenbacher *Melolontha furcicauda* Ancy *Eubrachys indica* Bally	Leaf eating beetles	Defoliate apple trees
Thysanoptera *Taeniothrips rhopalantinnalis* Shunister	Apple blossom thrips	Nymphs and adults scrap tissues from flower buds.
	(10) Pear (Pyrus communis)	
Diptera		
Dacus dorsalis (Hendel)	Fruit flies	See under mango
D.Zonatus Saunders		
Lepidoptera		
Carpocapsa pomonella Linnaeus	Codling moth	See under apple

TABLE 5.1 *(Contd.)*

Othreis fullonica Linnaeus	Fruit sucking moths	see under citrus
O. materna (Linnaeus)		
Hemiptera		
Quadraspidiotus perniciosus (Comstock)	San Jose scale	See under apple
Coleoptera		
Dorysathenes hugeli Redtenbacher	Apple root borer	See under apple
Lachnosterna coriacea (Hope) *Anomala dinidiata* (Hope) *Mimastra cyanura* (Hope)	Defoliating beetles	Adults feed on foliage
	(11) Peach (Amygdalus persica)	
Diptera		
Dacus dorsalis Hendel *D. zonatus* (Saunders)	Fruitflies	See under mango
Lepidoptera		
Anarsia lineatella Zeller	Peach twig borer	Caterpillars bore into twigs and feed on pulp and inner bark.
Calpe ophideroides Guenne	Fruit sucking moth	Moths suck the juice from fruits
Cacoecia epicyrta Meyrick	Peach leaf-roller	Overwintering larvae web around leaves, flower buds, blossoms and fruits and feed upon these in spring.
Hemiptera		
Pterochlorus persicae Cholodkovasky	Peach black aphid	Nymphs and adults suck sap from leaves, flowers, fruits and petioles
Myzus persicae (sulzer)	Green aphis	
Hyalopterus pruni (Geoffroy)	Peach mealy aphis	
Brachycaudus helichrysi (Kaltenbach)	Peach curl aphis	
Coleoptera		
Sphenoptera lafertei Thompson	Peach stem borer	Adults feed on leaves, grubs make galleries inside the bark.

TABLE 5.1 *(Contd.)*

Mimastra cyanura Hope	Swarming beetles.	Adults feed on leaves.
	(12) Plum (Prunus domesticus)	
Lepidoptera		
Agrotis spp.	Cutworm	Cut the tender nursery plants at ground level
Euproctis fraterna (Moore)	Plum hairy caterpillar	Young caterpillars feed gregariously on the epidermis of leaves; also feed on the epidermis of young fruits (Batra,1983).
Euproctis lunata Walker	Castor hairy caterpillar	Caterpillars feed gregariously on leaves.
Hemiptera		
Myzus persicae (Sulzer)	Green aphis	See under peach.
Hyalopterus pruni (Geofroy)	Peach mealy aphis	See under peach.
Brachycaudus helichrysi (Kalt)	Peach leaf curl aphis	See under peach.
Pterochlorus persicae Chlodkovasky	Peach black aphid	See under peach.
Quadriaspidiotus perniciosus (Comstock)	San Jose scale	Sucks sap from the succulent bark of stem, branch, twig, leaf stalk, fruit stalk and soft surface of leaf and fruit (Batra, 1983).
Eulecanium, coryli (Linnaeus)	Plum lecanid	Suck sap from stems, branches, shoots, leaves and fruits.
Coleoptera		
Sphenoptera lafertei Thomson	Peach stem borer	The larva feeds on the bark; when several larvae feed inside and girdle around the bark, the branch of whole tree may die (Batra, 1983).
Myllocerus undecim-pustulatus Faust	Grey weevil	Leaves are damaged severely (Batra, 1983).
	(13) Strawberry (Fragarice vesca)	
Lepidoptera		
Agrotis ipsilon Rott. *A. segetis* (Schiff)	Cutworms	Cut leaves at ground level.
Laphygma exigua (Hubner)	Lucerne caterpillar	Caterpillars are reported to bore into ripe fruits (Batra, 1983)

TABLE 5.1 *(Contd.)*

Euproctis fraterna (Moore)	Plum hairy caterpillar	See under plum
Coleoptera		
Holotrichia longipennis Blanchard	'Kurmula' beetle	Grubs feed on roots and adults on leaves.
Myllocerus undecimpustulatus Faust. *M. blandus* Faust	Grey weevils	Foliage is damaged severely by the weevils.
	(i) Vegetable crops *(1) Potato (Solanum tuberosum)*	
Lepidoptera		
Gnorimoschema (=Phthorimoea) operculella (Zeller) *G. heliopa* (Loer)	Potato tuber moth	Caterpillars of the moth cause damage by mining the leaves and also in exposed tubers in the fields, caterpillars also damage tubers of stored potatoes.
Agrotis ipsilon Rott. *A. flammatra* Sch. Muller	Cut worms	See under sorghum
Hemiptera		
Amrasca biguttula biguttula Ishida	Cotton jassid	See under cotton
Bemesia tabaci Geunadius	White fly	See under cotton
Aphis gossypii Glover	Cotton aphid	See under cotton (also responsible for spreading potato virus).
Myzus persicae Sulzer	Green aphid	See under mustard
Coleoptera		
Epilachna vigintiocto-punctata (=12 *punctata*) Mulcant *E. duodecastigma* Wied.	Epilachna beetles	Grubs and adults feed on leaves by scraping the chlorophyll and thus skeletonise the leaves.
Xylotrupes gideon Linnaeus *Lachnosterna longipennis* Blanchard	White grubs	Grubs feed on the root and kill the plant
	(2) Tomato (Lycopersicum esulentum)	
Lepidotera		
Heliothis armigera Hubner	American bollworm	Caterpillars bore fruits and feed inside on the pulp

TABLE 5.1 *(Contd.)*

Spodoptera litura Fabricius	Tobacco caterpillar	See under tobacco
Coleoptera		
Epilachna spp.		See under potato
Hemiptera		
Amrasca biguttula biguttula Ishida	Cotton Jassid	See under cotton
Bemisia tabaci Gennadius	Cotton White fly	See under cotton (transmit the leaf-curl disease)
Trialeurodes vaporarium Westwood	White fly	Adults and nymphs suck sap from lower surface of leaves
Pseudococcus virgatis Cockerell	Scale insect	By sucking sap from tender shoots and leaves.

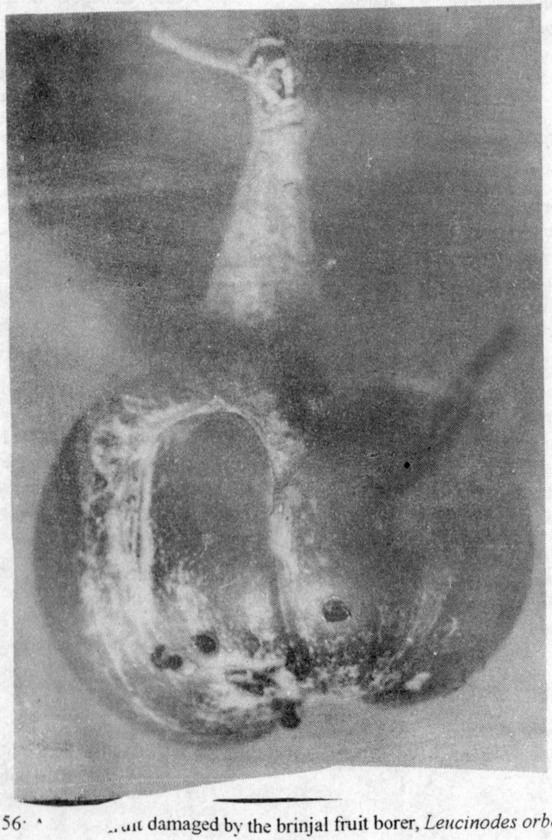

FIG. 56· ˈ ..uit damaged by the brinjal fruit borer, *Leucinodes orbonalis*

TABLE 5.1 *(Contd.)*

	(3) Brinjal (Solanum melongena)	
Lepidoptera		
Leucinodes orbonalis Guenee	Brinjal shoot and fruit borer	By boring into shoots and fruits (Fig.56)
Euzophera perticella Ragonot	Brinjal stem borer	By boring into stem.
Scrobipalpa blapsigona M.	Bud worm	By boring into flowers and buds.
Eublemma olivacea Walker	Leaf caterpillar	By rolling and feeding on leaf chlorophyll
Hemiptera		
Amrasca biguttula biguttula Ishida	Cotton jassid	By sucking cell sap and transmitting little leaf and mosaic diseases.
Urentius hystricellus Richt	Brinjal lace wing bug	Nymphs and adults suck sap from the underside of leaves (Prasad *et al.* 1983)
Aphis gossypii Glover	Cotton aphid	See under cotton.
Coleoptera		
Epilachna duodecastigma Wied. *E.vigintioctopunctata* Fabricius	Epilachna beetles	See under potato
	(4) Okra (Abelmoschus esculentus)	
Lepidoptera		
Earias insulana Boisduval *E. vittella* Fabricius	Spotted boll worms (Fig.57)	See under cotton
Sylepta derogata Fabricius	Leaf roller	See under cotton
Coleoptera		
Myllocerus spp.	Grey weevil	See under cotton
Hemiptera		
Dysdercus Koenoigii Fabricius	Red cotton bug	**See under cotton**
Amrasca biguttula biguttula (Ishida)	Cotton Jassid	
Bemisia tabaci Gennadius	Cotton Whitefly	**See under cotton**

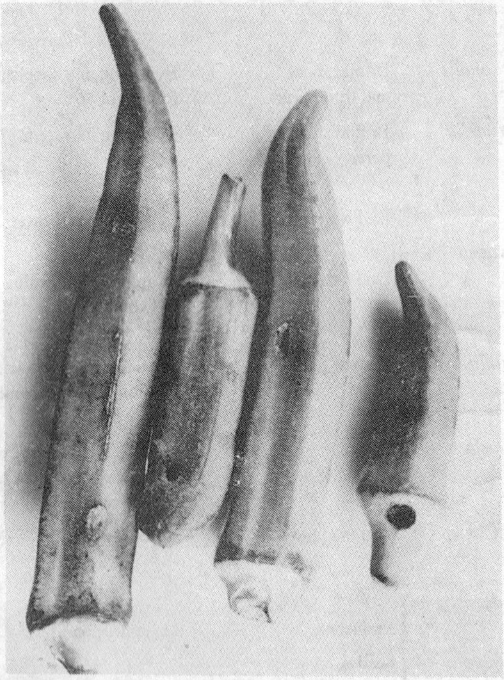

FIG. 57: Okra fruits damaged by the spotted boll worm *Earias vittella*

TABLE 5.1 *(Contd.)*

	(5) Beans *Phaseolus vulgaris, Dolichus lablab etc.)*	
Diptera		
Ophiomyia (Melanagromyza) phaseoli Tryon	Bean stem-fly	See under pulses.
Lepidoptera		
Heliothis armigera Hubner	American bollworm (Gram caterpillar)	See under tomato
Aphis craccivora Koch.	Bean aphid	Sucks sap from leaves, tender shoots, buds and flowers, spreads virus and mosaic diseases (see also groundnut).

TABLE 5.1 *(Contd.)*

	(6) Peas (Pisum sativum)	
Diptera		
Phytomyza atricornis Meegen	Leaf miner	By mining in leaves by maggots.
Melanagromyza phaseoli Coquillett.	Pea stem fly	See under green gram
Lepidoptera		
Etiella zinckenella Treitschke	Pod borer	By making hole in pods and feeding on seeds.
Laphygma exigua Hubner	Lucerne caterpillar	See under Bengal gram
Heliothis sp. Fabricius		By boring into pods and feeding on seeds.
Hemiptera		
Acyrthosiphum pisi Harris	Aphid	By sucking sap from leaves and tender shoots by nymphs and adults.
Thysanoptera		
Caliothrips indicus Bagnall	Thrips	By lacerating leaves and tender growing parts by nymphs and adults.
	(7) Cucurbitaceous vegetables	
	Pumpkin *(Cucurbita pepo)*, bottle gourd *(Lagenaria leucantha)* Sponge gourd and Ridgegourd (*Luffa* spp.), Bittergourd *(Memordica charantia)*, Cucumber *(Cucumis sativus)*, Snakegourd *(Trichasanthes anguina)* etc.	
Diptera		
Dacus cucurbitae (Coquillett) *D. ciliatus* (loew) *D. latifrons* (Hendel) *D. diverses* (Coquillett) *D. parvulus* (Hendel)	Fruit flies	Females lay eggs under the epidermis of the fruit by piercing it (epidermis) and the hatching magots get inside the fruits and feed on the pulp (Fig. 58)
Lepidoptera		
Margaronia indica (Saunders)	Pumpkin caterpillar	Caterpillars feed on the leaves of the creepers.
Sphenarches caffer Zeller	plume moth	
Plusia peponis (Fabricius)	Snake-gourd semilooper	Caterpillars feed on leaves.

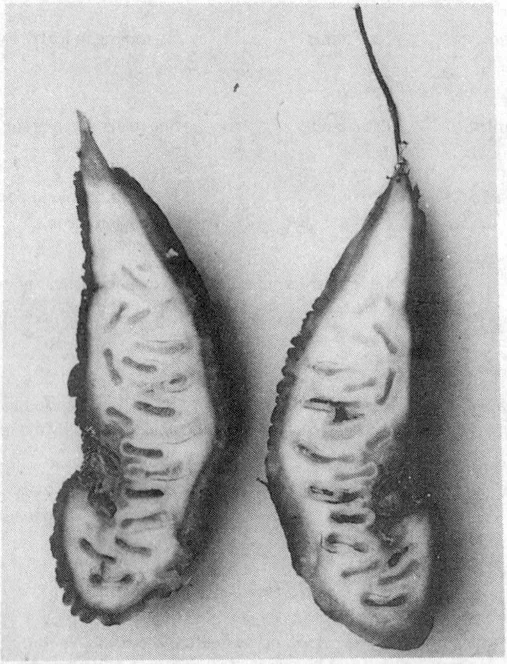

FIG. 58: A bitter gourd fruit cut open to show the damage caused by the fruit fly, *Dacus cucurbitae* grubs

TABLE 5.1 *(Contd.)*

Hemiptera		
Aspongopus janus Fabricius	Stink bug	Nymphs and adults suck sap from leaves and tendershoots
Aphis gossypii Glover	Cotton aphid	See under cotton
Coleoptera		
Raphidopalpa foveicollis Lucas	Pumpkin beetles	Grubs damage roots and adults feed on flowers, buds and leaves.
R. intermedia Jacoby		
R. cincta Fabricius		
Epilachna duodecastigma Mulsant	Epilachna beetles	See under potato
E vigintioctopunctata Fabricius		

TABLE 5.1 *(Contd.)*

Mylabris pustulata Thunberg	Blister beetle	Adults feed on pollen and petals of flowers

(8) Cruciferous vegetables

Cabbage *(Brassica oleracea* var. *capitata),* Cauliflower *(B. oleracea* var. *botrytis),* knol-khol *(B. oleracea* var. *gongylodes),* turnip *(B. campestris* var. *rapifera),* Radish *(Raphanus sativus),* Chinese cabbage *(B. pekinensis)* and Brussels sprouts *(B. oleracea* var. *gemmifera).*

Hymenoptera		
Athalia lugens proxima Klug.	Mustard sawfly	Grubs feed by making holes in leaves; mainly pest of nursery/young plants.
Lepidoptera		
Plutella maculipennis Curtis	Diamond back moth	The leaves may be completely skeletonised by the caterpillars.
Pieris brassicae Linnaeus	Cabbage butterfly	See under mustard
Helula undalis Fabricius	Cabbage-borer	By mining the foliage, and feeding on shoot and finally boring into the stem.
Crocidolomia binotalis Zeller	Cabbage leaf webber	By webbing and feeding on leaves by caterpillars.
Plusia orichalcea (Fabricius)	Cabbage semilooper	By feeding on foliage by caterpillars.
Hemiptera		
Bagrada cruciferarum (Linnaeus)	Painted bug	Nymphs and adults suck sap from tender leaves and shoots.
Brevicoryne brassicae Linnaeus *Lipaphis erysimi* Kaltenbach	Mustard aphids	Suck sap from leaves and shoots, secrete honey dew on which sooty mould grows
Myzus persicae Sulzer	Green peach aphid	Acts as a carrier of about 50 viral and mosaic diseases.

(j) Ornamentals

(1) Rose (Rosa spp.)

Hemiptera		
Macrosiphum euphoebia Thomas *M. rosae* Linnaeus	Aphids	Both nymphs and adults cause damage by sucking cell sap from different parts of plants. Due to

TABLE 5.1 *(Contd.)*

		aphid attack, tender shoots wither fall prematurely and flowers lose the beauty (Butani, 1974).
Aleurocanthus spp.	White fly	Both nymphs and adults suck the cell sap from plants.
Aonidiella aurantii Maskell	The red scale	Cover the tender shoots, suck the cell sap and in severe cases of infestation, the shoot/plant may even die.
Aspidiolus lataniae Signoret *A. orientalis* Newstead *Icerya aegyptiaca* Douglas	Scale insects	Their mode of damage is similar to that of red scale but their incidence is not very common.
Thysanoptera *Rhipiphorothrips cruentatus* Hood	Thrips on leaves	These insects cause damage by rasping the tissues and sucking the sap that oozes out of the wounds

FIG. 59: Cockchafer beetle, *Adoretus* spp. defoliating a rose twig.

TABLE 5.1 *(Contd.)*

Retithrips syriacus (Mayet)	(Ananthakrishnan, 1971)	caused by rasping leave brown patches on leaves, flowers and petals.
Thrips coloratus Prienser *T. florum* Schmutz	Thrips on flowers (Ananthakrishnan, 1971)	The attacked leaves and flowers finally wither (Butani, 1974).
Coleoptera *Adoretus versutus* Harold *A. Cardoni* Brenske *A. caliginosus* Burmeister *A. lasiophygus* Burmeister *A. bangalorensis* Brenske	Cockchafer beetles	Grubs and adults both causes damage, grubs feed on roots and adults on leaves voraciously by making irregularcuts on leaves. The entire plant may be almost defoliated (Fig. 59).
Lepidoptera		
Ascotis imparata Walker		Caterpillars feed on leaves.
	(2) Chrysanthemum (Chrysanthemum spp.)	
Thysanoptera *Fraukliniella daupfi* Priesner	Thrips	Both nymphs and adults cause damage by lacerating the soft tissues on the neutral surface of leaves and feed on the cell sap oozing out of the wounds caused by lacerating as a result the leaves gradually start drying.
	(3) Jasmine (Jasminum arborescens)	
Hemiptera *Dialeurodes spp.* *Bemisia giffardi* (Kotinsky)	White flies	Both nymphs and adults suck cell sap especially of leaves and petals.
Thysanoptera *Thrips hawaiensis* (Morgan) *Isothrips orientalis* Bagnall	Thrips	Both nymphs and adults lacerate the soft tissues on the ventral surface of leaves and feed on the cell sap oozing out of these wounds; such leaves slowly dry.
	(4) Sunflower (Helianthus annus)	
Hemiptera *Aphis gossypii* Glover	Aphid	Both adults and nymphs suck cell sap from different parts of plants (Butani and Dadlani, 1983)

TABLE 5.1 *(Contd.)*

Thysaoptera *Scitothrips dorsalis* Hood *Fraukliniella daupfi* Priesner	Thrips	Both nymphs and adults lacerate the soft tissues on the ventral surface of leaves and feed on the cell sap oozing out of these wounds; such leaves slowly dry (Saxena, 1987)
	(5) Gladiolus (Gladiolus L.)	
Hemiptera *Aphis craccivora* koch,	Bean aphid	Both adults and nymphs suck cell sap from different parts of plants.
Thysanoptera *Taeniothrips simplex* (M)	Thrips	Both nymphs and adults lacerate the soft tissues on the ventral surface of leaves and feed on the cell sap oozing out of these wounds; such leaves slowly dry (Saxena, 1987).
	(6) Money plant (Scindapsus (Pothos aureus)	
Thysanoptera *Prignothrips medanensis* Priesener	Thrips	Both nymphs and adults lacerate the soft tissues on the ventral surface of leaves and feed on the cell sap oozing out of these wounds; such leaves slowly dry up (Saxena, 1987).
	B. INSECT PESTS OF STORED GRAINS AND OTHER COMMODITIES	
Lepidoptera *Corcyra cephalonica* Stainton	Rice moth	Caterpillars feed on all kinds of cereals and web broken grains into a mass (external feeder). infested grains rendered unfit for human consumption, occasionally damage dry fruits groundnut etc.
Ephestia (Cadra) cautella Walker	Fig/Almond moth	Caterpillars feed on all kinds of cereals (external feeder) and cereal products, fruits, condiments, cotton seeds, dates and other nuts etc.
Sitotroga cerealella Olivier	Angoumois grain moth	Caterpillars bore into the grain cereals and feed (internal feeder).

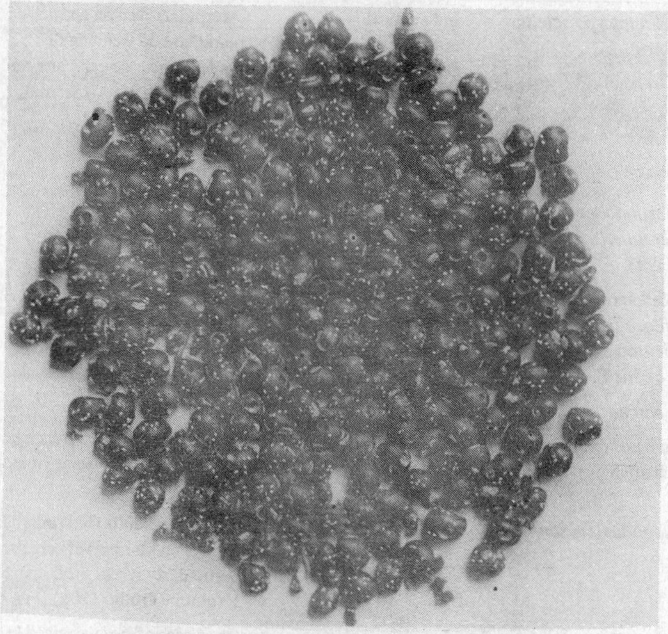

FIG. 60: Black gram grains damaged by the pulse beetle, *Callosobruchus maculatus*

FIG. 61: Wheat grains damaged by the rice weevil, *Sitophilus oryzae*

TABLE 5.1 *(Contd.)*

Plodia interpunctella Hubner	Indian meal moth	Caterpillars feed on germ point and do webbing of whole grains, (cereals) and milled cereals into a mass (external feeder); occasionally feeding on dry fruits (Pruthi, 1969).
Gnorimochema opercullella (Zeller)	Potato tuber moth	See under potato
Coleoptera		
Callosobruchus maculatus F. *C. analis* F. *C. chinensis* L.	Pulse beetles	Grubs feed on whole pulses by boring (internal feeder) (Fig. 60)
Zabrotes subfaciatus Boheman	Red kidney bean beatle/bean beetle	Grubs feed on stored french bean, cowpea and other beans (internal feeder).
Sitophilus oryzae L.	Rice weevil	Grubs and adults feed (internal feeder) on all kinds of cereals, particularly wheat, rice, maize and sorghum (Pruthi, 1969) (Fig. 61)
Rhizopertha dominica F.	Lesser grain borer	As in case of rice weevil
Lasioderma serricorne F.	Tobacco beetle/ cigarette beetle	Grubs and adults feed (internal feeder) on all kinds of spices and also bore in dry tobacco e.g. cigarette cigar, etc. (Ayyar, 1940). Even cereals are not spared.
Stegobium panicium L.	Drug store beetles	Grubs and adults feed on all kinds of cereals, also found boring in dry spices like turmeric, ginger, coriander (Ayyar, 1940), black and red pepper (Pruthi, 1969). Also attack wheat flour.
Trogoderma granarium Everts	Khapra beetle	Only grubs feed, start feeding on germ point (external feeder) of cereals and continue feeding deeper. Also feed on broken pulses and dry fruits.
Oryzaephilus surnamensis Linnaeus	Saw-toothed grain beetle	Grubs feed externally on grains, dried fruits, biscuits, etc.
Latheticus oryzae Waterhouse	Long headed flour beetle	Grubs and adults feed externally on cereals flour.

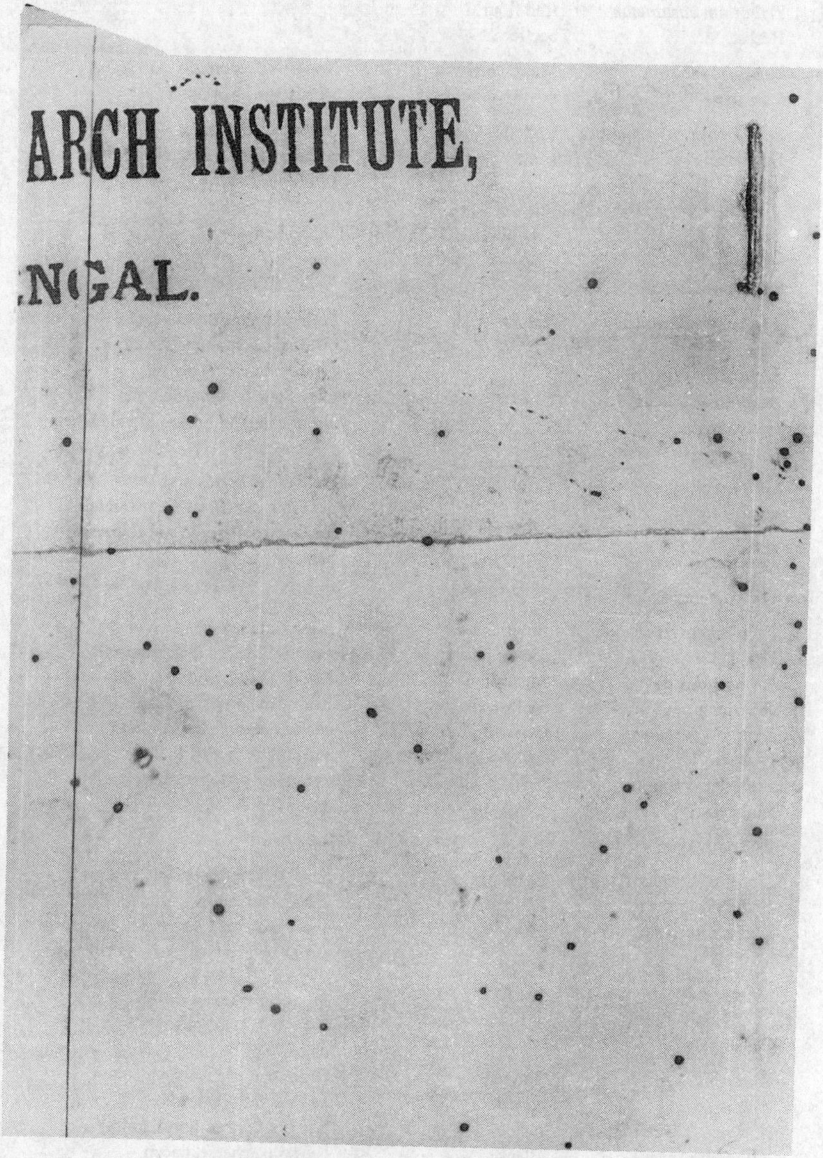

FIG. 62: Paper damaged by the silver fish, *Lepisma* sp.

TABLE 5.1 *(Contd.)*

Tribolium castaneum Herbst.	Red flour beetle	
Tenebrio molitor Linnaeus	Meal worm	Grubs feed externally on cereals and its products.
Tenebrioides mauritanicus Linnaeus	Cadelle beetle	Grubs feed externally on cereals, only occasionally (Pruthi 1969).
	C. HOUSEHOLD INSECT PESTS	
Thysanura		
Lepisma saccharina Linnaeus *L. nigrina Silvestri* *Acrotelsa collaris* (Fabricius)	Silver fish	Feed on vegetable and cereal materials, containing starch; damage books for eating glue. Sometimes make holes also in pages of books (Fig. 62)
Orthoptera		
Gryllodes sigillatus	Cricket	Food range varies from cereals to decaying matter, nymphs and adults, feed on woollen and silk fabrics, books etc.
Dictyoptera		
Periplaneta americana Linneus *P. australasiae* (Fabricius) *Blattella germanica* Linnaeus *Blatlella orientalis* Linnaeus	American cockroach Australian cockroach German cockroach	Adults and nymphs feed on various food products especially cereals, cheese and starchy materials, damage books, consume human food and pollute food with their excreta by giving it roachy odour (Pruthi,1969).
Psocoptera		
Liposcellis transvalensis Enderlein *L. divinatorius* (Mueller)	Book lice or cereal psocid	Feed on cereal products, vegetable and animal debris, paste, glue, dried, preserved insects sometimes seriously infest tea packed in chests and stored in damp godowns (Pruthi, 1969)
Coleoptera		
Anthrenus scrophulariae Riley *A. flavipes* Waterhouse *Attagenus piceus* (Olivier) *A. gloriosae* Fabricius *Dermestes vulpinus* Fabricius	Spotted carpet beetles Black carpet beetles Carpet beetle	Grubs damage woollen, cotton and silk fabrics, leather, brushes, feather, horns, hides, bees wax, cheese, bacon, museum specimens of insects etc. (Pruthi,1969). (Fig.63)

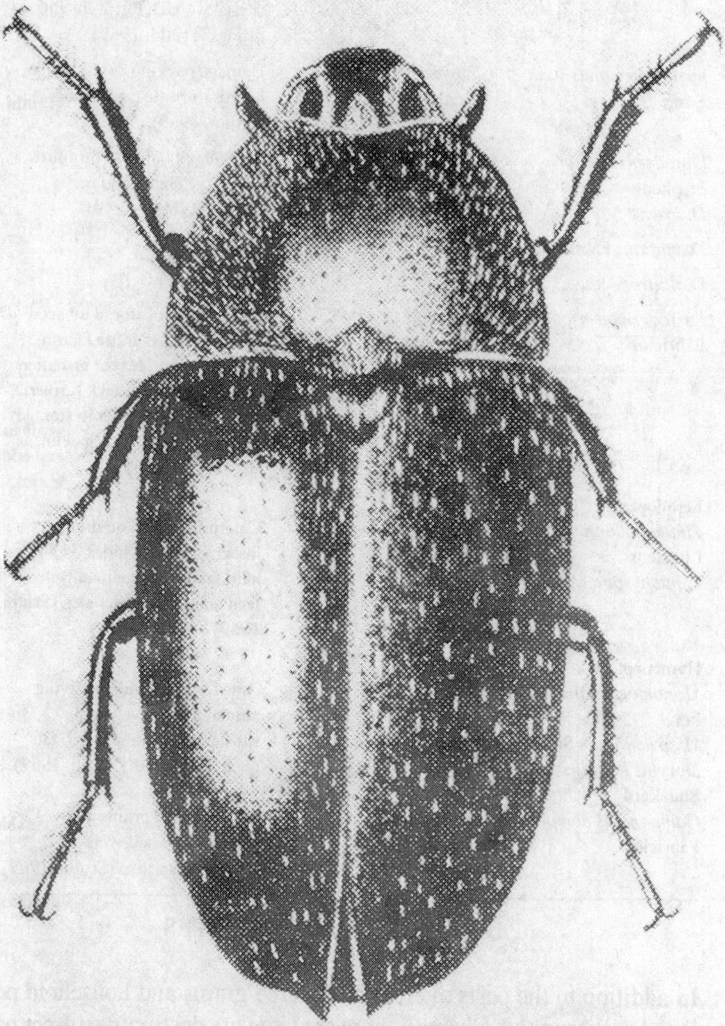

FIG. 63: The carpet beettle, *Dermestes vulpinus*

TABLE 5.1 *(Contd.)*

Heterobostrychus aequalis Waterhouse	Furniture beetle	Grubs and adults bore in packing cases, plywood, chests, wooden furniture and fittings (Pruthi, 1969).
Lyctus africanus Lesne	Powder post beetle	Damage packing cases, planks, boards, furniture, plywood, etc. (Pruthi, 1969).
Dinoderus ocellaris Stephens *D. brevis*	Shoot borers or shot-hole	Also attack bamboo furniture e.g., large chairs and racks (Pruthi, 1969).
D. *minutus* Fabricius	borers	
D. *pilifrons* lesne		
Lasioderma sericorne Fabricius	Cigarette/ tobacco/ Cheroot beetle	Serious pest of stored tobacco in cigar and cigarette factories, attack tobacco leaves just after harvesting, also attack turmeric, black and red pepper in storage; preserved, dry, plant specimens etc. (Pruthi, 1969).
Lepidoptera		
Tinaea pellionella Linnaeus *T. pachyspila* Meyrick	Clothes/fur moths	Caterpillars feed on dried insect specimen, hides, wool, hair, feather, occasionally feed on milk powder also (Pruthi, 1969).
Hymenoptera		
Monomorium indicum Forel *M. gracillinum* Smith *Dorylus labiatus* Shuckard	Ants	Ants feed on dead and living animal matter, vegetable substances, grains, fungi, fats, meat and sweets (Pruthi, 1969).
Camponotus compressus Fabricius	Carpenter ant	Attack stored grains and sometimes wood work of old buildings also (Pruthi, 1969).

D. LOCUSTS AND TERMITES

In addition to the pests of crops and stored grains and household pests listed earlier in this Chapter, there are two very destructive insect pests which cause serious damage to practically all the crops. These two pests are locusts and termites.

Locusts are considered to be the oldest winged-enemy of mankind. These are polyphagous pests causing damage to crops in several countries of the world. Several species of locusts are known e.g.,

(1) *Locusta migratoria migratoides* Linnaeus of Tropical Africa.
(2) *L. paradalina* Walker of South America.
(3) *Schistocera paravensis* Burmiester of South America
(4) *Deciostames moracanus* Thunberg of the Mediterranean region.
(5) *Nomadacris septemfasciata* Serville of the African region South of Equator.
(6) *Patanga succincta* Linnaeus, the Bombay Locust
(7) *Anacridium rubrispinum* Bei-Benko, the Tree Locust
(8) *Locusta migratoria* Linnaeus, the Migratory Locust
(9) *Schistocerca gregaria* Forskal, the Desert Locust

The last 4 (i e. 6,7,8 and 9) are found in the Indian region. Out of these four species *Schistocerca gregaria,* the Desert locust is most important. It affects 60 countries of the world including India and its neighbouring countries. It feeds upon practically all the plants. Some of its preferred food plants are: cabbage, cauliflower, maize, sorghum, cotton, guava, wheat and peas etc. When a swarm happens to settle on trees, it consumes their bark also. The only plants not consumed by this pest are: *Azadirachta indica, (syn. Melia azadirachta,* and *Calotropis* sp. So far there is only one record of desert locust swarm simply nibbling *A. indica* (neem) leaves when the swarm settled on trees. This is a very exceptional case and the locust appears to have nibbled the leaves because of being very hungry. When the adults become sexually mature, their feeding is reduced, mating starts and the mated females lay eggs after locating suitable places for oviposition.

In India, Jaisalmer, Barmer and Bikaner in Rajasthan are the permanent breeding grounds of Desert Locust. Besides, locust enters India from Iran, Iraq, Afghanistan and Pakistan. If locust is controlled in the breeding ground itself, no swarm formation will take place. If it happens to be otherwise, the population in the breeding ground goes on increasing. When the population of adults becomes 18,000-20,000 per sq km, the swarm formation takes place as there is shortage of food and space for egg laying. Swarm leaves the breeding ground and settles on the crops wherever the sun sets and it becomes dark. If the swarm happens to be a pink one i.e., sexually immature, the crop on which it settles is completely eaten up; it happens to be a yellow one i.e., sexually mature, it does not cause much damage to the crops by feeding, but there is heavy egg-laying and the hatching-hoppers may cause very serious

FIG. 64: The desert locust, *Schistocerca gregaria* congregating on a tree and defoliating it

damage to the nearby crops, if they are not controlled in time (Fig.64)

Termites may be considered to be more destructive than Locusts. While locusts are only crop pests, termites, in addition to causing severe damage to agricultural/horticultural/plantation crops and forest trees, also attack wooden fittings, furniture, constructional timber, bamboo work, paper, books and textiles etc.(Fig.65) As a matter of fact termites may feed on any material containing cellulose.

Some of the agricultural/horticultural crops attacked by termites are: paddy, barley, maize, pearl millet, pulse crops, sugarcane, cotton, jute, groundnut, mango, guava, citrus, sapota, chillies, potato, cabbage and cauliflower etc. Termites also attack spices e.g. cumin, aromatic plants like Japanese mint and tea, Coconut and rubber plantations, *Eucalyptus* nurseries and *Azadirachta indica* trees (Roonwal,1983)

About 1900 species of termites are known from the world out of which, 260 have been recorded from India. Of the Indian 25 species (Roonwal, 1983) listed below have been reported causing damage to agricultural/horticultural/crops and forest trees/saplings:

Glyptotermes dilatatus (Bug. & Pop.), *Bifiditermes beesoni* (Gard.),*Anacanthotermes macrocephalus* (Desn.), *A. viarum* (Koenig), *Heterotermes indicola* (Wasm.), *Coptotermes*

FIG. 65: A piece of wood damaged by termites

ceyloicus Holmg. *C. heimi* (Wasm.), *Ermotermes paradoxalis* Holmg., *Ermotermes* sp., *Microcerotermes beesoni* Snyd., *M. minor* Holmg., *Microcerotermes* sp., *Dicuspiditermes fletcheri* (Holmg. & Holmg.). *Odontotermes assmuthi* Holmg., *O. feae* (Wasm.), *O. obesus* (Ramb.), *O.parvidens* Holmg.& Holmg.,*O.redemanni* (Wasm.), *O. taprobanes* (Oalk.), *O. wallonensis* (Wasm.), *Microtermes obesi* Holmg.,

Nasutitermes sp., *Trinervitermes biformis* (Wasm.), *T. indicus* (Snyd.), *and T. rubidus* (Hag.).

Termites are social insects and live in colonies. The colony may sometimes be as big as one containing over a million individuals (Roonwal, 1983). The colonies consist of individuals of different castes e.g., soldiers, workers, functional males and females (queen). They live mostly underground and come up for foraging. Normally roots of young plants/saplings are attacked first. Later termites extend their activity to stem which is hollowed out. In sugarcane, the eye-buds of the setts are attacked after sowing and later the damage extends to stem. In older trees, the trunk is attacked and in severe infestation, the tree may die unless control measures are taken in time (Roonwal, 1983).

E. INSECT VECTORS OF PLANT VIRUSES

Vector	Virus	Method of transmission
Orthoptera		
Leptophyes punctatissima	Turnip yellow mosaic	Virus-transmission by contamination of mouth-parts of the insects with virus during feeding on diseased leaves and then healthy leaves.
Tettigonia viridissima	Potato virus x, Tobacco mosaic	
Melanoplus differentialis	Potato virus x	
Dermaptera		
Forficula auricularia	Turnip yellow mosaic	Mechanical transmission like beetles (see under Coleoptera)
Thysanoptera		
Frankliniella insularis	Tomato spotted wilt	Thrips can only transmit viruses by acquiring them during nymphal stage.
Thrips tabaci	Sunflower mosaic	
Coleoptera		
Acalymma trivittata	Squash mosaic	Mechanical transmission; transmit viruses during feeding and regurgitating the infected chewed leaves.
Ceratoma ruficornis	Cowpea mosaic	
Phyllotreta spp.	Turnip yellow mosaic, turnip crinkle	
Systena toeniata	Potato spindle tuber	
Lepidopetra		
Heliothis assulta	Tobacco mosaic	

Vector	Virus	Method of transmission
Diptera		
Liriomyza longii	Tobacco mosaic	Mechanical transmission
Phytomyza rufipes	Turnip crinkle	—mouth part contamination
Hemiptera		
Nysius sp.	Centrosema mosaic	Mechanical transmission;
Perkinsiella saccharicida	Sugarcane fiji	Viruses are generally deep-seated in tissue
P. vastiatrix	Tomato big bud	and require longer
Hyalesthes obsoletus	cransberry false blossom virus	feeding by the insects for transmission. They are not stylet borne.
Aceratagallia sangiunolenta	Potato yellow dwarf	These are phloem, Xylem and Parenchyma
Agallia quadripunctata		feeders but transmit virus efficiently
Euttetix tenellus	Beet curly top virus	when feed in phloem tissues for longer period.
Draeculacephala portiala	Sugarcane chlorotic streak, lucern dwarf	
Perigrinus maidis	Maize mosaic virus	Viruses need incubation period as
Colladonus geminatus	Aster yellows	they are of propogative type and multiply in the vectors. In some cases
Baldulus maidis	Maize stunt	transovarial transmisson also occurs.
Macrosteles divisus	Aster yellows	
Nephotettix cincticeps	Rice dwarf	
Orosius argentatus	Groundnut mosaic	
Bemisia tabaci	Abutilon infectious varigation, double bean yellow mosaic, cassara mosaic, cotton leaf curl, *Dolichos lab lab* yellow mosac, pumpkin yellow mosac, sweet potato stunt, tobacco leaf curl.	Mostly phloem feeder and transmit by feeding on diseasesd and healthy plants. Viruses need incubation period. Viruliferous white flies remain infective throughout life; more than one virus can be transmitted by a whitefly at a time.
Ampharophora rubi	Raspberry yellow blotch virus	Stylet borne viruses adhere to the stylets during probings and are
Aphis nasturtii	Cucumber mosac Potato virus X, Potato virus A, Potato leaf roll.	transmitted. mechanically. Viruses are located in superficial tissues.
Aphis fabae	More than 35 viruses of bean, sugarbeet, cowpea, potato and legumes etc.	

Vector	Virus	Method of transmission
Aphis gossypii	More than 44 viruses of ornamentals, legumes and vegetables,etc.	Circulative viruses are swallowed by aphids and have to pass through the gut wall into the blood and back to salivary glands, before the alphids become infective; viruses are generally located in phloem tissues. They are retained by insects for longer period.
Brevicoryne brassicae	More than 15 viruses of cruciferous plants.	
Myzus persicae	More than 99 viruses of different crops and plants.	Propogative-viruses, have definite biological relationship with their vectors as they multiply inside the vector e.g., Potato leaf roll, sowthistle yellow-vein mosaic virus, and strawberry crinkle etc.
Myzus spp. (other than *Myzus pessicae)*	More than 21 viruses of different plants.	
Neomyzus circumflexus	More than 30 different viruses.	
Macrosiphum solanifolii	More than 50 different viruses	Prologed feeding for virus acquisition.
Pseudococcus njalensis	Cocoa swollen shoot	

F. INSECTS AS ANIMAL PESTS/DISEASE VECTORS

(1) Cattle/Buffaloes

Scientific name	Common name	Nature of damage/disease
Diptera *Tabanus striatus* *T. rubidus* *T. tenens* *T. virgo* *T. ditaeniatus* *Chrysops dispar*	Horseflies (Tabanids)	Only female tabanids bite and suck blood from cattle, buffaloes, horses, camels and elephants (Mohan,1981). Tabanid species referred to transmit causative organisms of "Surra" disease (Pruthi,1969).
Haematopota javana		
Phlebotomus argentipes *P. sergenti* *P. papatasii*	Sandflies	Like horseflies, in sandflies also *only* females suck blood from their animals hosts. *P. argentipes* vector of Kala-azar shows preference for cattle blood. *P. sergenti* is vector of cutaneous leishmaniasis (Mohan, 1981). *P. papatasii* is well known for transmitting the sandfly fever of man (Pruthi,1969).
Culicoides oxystoma *C. anophlis* *C. peregrinus*	Midges	Only the female midges bite, especially in the evening these attack animals in large numbers and inflict painful bites (Mohan,1981). Also act as intermediate host of fillaria

Scientific name	Common name	Nature of damage/disease
		worma like *Onchocera* in cattle and horses.
Simulium indicum *S. grisescens* *S. rufibasis* *S. rufithorax* *S. striatum*	Buffalo gnats (Black flies)	Females suck blood by biting, may enter orifices of ears to suck blood (Pruthi, 1969). Attack cattle, buffaloes and horses in swarms, cause great annoyance, sometimes even death within a few hours due to loss of blood as well as toxaemia. Certain species transmit onchocerciasis' in cattle (Mohan, 1981).
Aedes aegypti	Mosquito	Female mosquitoes suck blood of animals.
Siphunculina funicola	Eye-fly	Hover in front of eyes and feed on secretion of eyes (Pruthi, 1969).
Musca vicina *M. nebulo* *M. domestica*	Houseflies	Act as carriers of diseases. Also carry mastitis in cattle. House fly is also intermediate host of stomach worm of horse (Pruthi, 1969).
Musca crassirostris	Blood-sucking fly of cattle	Several often seen obtaining their fill of blood avidly from the same animal (Pruthi, 1969).
Hypoderma lineata *H. bovis*	Warble-fly/ Common cattle grubfly	*H. lineata* common in India, attacks cattle, occasionally buffaloes also. Grubs hatching from the eggs laid on the hairs of legs bore through the skin and finally migrate to the back appearing as warble tumours. Full-grown larvae squeeze out for pupation through exit holes. Hides, while tanning found to have upto even 300 holes reducing its economic value.
Hippobosca maculata	Horse fly	Parasite of warm-blooded animals especially horses. Found on cattle buffaloes, pigs and camels also. Suck blood of the host animal; suspected to transmit *Trypanosoma theileri*. Some times become serious pest due to large quantity of blood sucked leading to the host's death (Mohan, 1981).

Scientific name	Common name	Nature of damage/disease
Siphonal (= Lyperosia) exigua	Horn fly	Both male and female flies suck blood of livestock and transmit the helminth *Habronema microstoma* and is perhaps vector of other diseases also.
Stomoxys calcitrans	Stable-fly	
Calliphora erythrocephala	European blue bottle	The grubs burrow into the tissues underlying the skin and thus cause myiasis in livestock.
C. vomitoria		
Lucilia cuprina	Green bottle	
L. sericata		
Chrysomyia bezziana	Screw-wormfly	
Sarcophaga ruficornis	Flesh flies	
S. haemorrhoidalis		
Siphunculata (Sucking lice)		
Haematopinus eurysternus	Short-nosed lice (on cattle)	Suck blood, cause irritation and milk production is reduced.
H. tuberculatus	(on buffaloes)	
Linognathus vituli	Blue lice (on calves)	
Solenoptes capillatus	(on cattle and buffaloes)	
Mallophaga	Biting lice	
Damalinia (Bovicola) bovis (Linnaeus)	Cattle biting louse (little red louse)	Bite, cause irritation by crawling on skin.
	(2) Sheep and Goats	
Siphunculata (sucking lice)		
Linognathus stenopsis (Bumeister)	Blue louse (on goat)	By sucking blood
L. pedalis	(on sheep)	
Mallophaga (Biting lice)		
Damalinia (Bovicola) ovis	(on sheep)	By feeding on exuviae scab etc. and causing irritation.
D. caprae	(on goat)	
Diptera		
Malophagus ovinus	Tick-fly or 'sheep Red' (on sheep)	Grubs and adults both suck blood from sheep, thus irritation is caused

Scientific name	Common name	Nature of damage/disease
		leading to rubbing of body and it results in loss of wool.
Hypoderma crossii	Warble-fly of goat	As in case of warble fly of cattle
Oestrus ovis	Sheep bot or head maggots of sheep.	Grubs infest nostrils and head sinuses which may result in death of sheep and goats.
	(3) Poultry	
Diptera		
Aedes spp.	Mosquito	By transmitting *Plasmodium gallinaceum* leading to paralysis and death (only experimentally proved). Natural vector is still unknown (Mohan, 1981)
Hemiptera		
Cimex hemipterus	Bed bug (of tropical countries)	It infests poultry houses. Adults and nymphs both suck blood from poultry. Some times attack domestic animals also (Mohan, 1981).
Siphonaptera		
Echidnophaga gallinacea	Stick-tight flea	Feed by embedding their head in the skin and often killing fowls. Occasionally parasitic on domestic animals especially dogs (Mohan, 1981).
Mallophaga (Biting lice)		
Eomenacanthus stramineus	Body louse or Vent iouse	Body louse is a serious pest of fowls found especially around the vent and breast.
Goniodes gigas	Large body louse and	Large body louse and shaft louse occur in body feathers.
Menopon gallinae	Shaft louse	
Gallilipeurus heterographus	Head louse	Head louse is found on the skin and feathers of head and neck.
Lipeurus caponis	Wing louse	Wing louse is found on the underside of large wing feather.
L. tropicalis	Tropical wing louse and	The Tropical wing louse and Fluff louse occur in the fluff at the base of feathers.
Goniocotes gallinae	Fluff louse	
		All the biting lice are ectoparasitic. They do not

Scientific name	Common name	Nature of damage/disease
		suck blood but have biting mouth parts and feed on scales, and base of feathers. By their irritation fowls become weak leading to reduction in egg production. Young chicken may often die due to their infestation.

G. INSECTS AS HUMAN DISEASE CARRIERS

Scientific name	Common name	Mode of infection
Diptera		
Musca nebulo	Housefly	Accidentally carry pathogens of cholera, dysentery, intestinal worms, leprosy, typhoid etc., from filth to food.
Aedes aegypti *A. leucocelaenus* *Hemagogus capricornii* *H. equinus* *H. albomaculatus*	Mosquitoes	By inoculating yellow fever viruses into blood.
Anopheles culicifacies *A. fluviatitis,* *A. stephensi* *A. minimus* and five other species	Mosquitoes	Inoculate into blood, the protozoan parasites, *Plasmodium vivax, P. falciparum* and *P. ovale*—the causative organisms of malaria.
Aedes aegypti *A. albopictus*	Mosquitoes	Inoculate dengue fever viruses into blood.
Culex tarsalis *C. pipens* *C. quinquefasciatus* *Aedes taeniorhynchus* *A. sollicitans* and others.	Mosquitoes	Inoculate encephalitic viruses into blood.
Several species of genera *Aedes, Anopheles, Culex* and *Mansonia*	Mosquitoes	Inoculate the nematode *Wuchereia bencrofi* and *W. malayi,* causative organisms of filariasis into blood.
Glossina morsitans others	Tsetse fly	Inoculate the protozoan parasites, *Trypanosoma gambienee* and *T. rhodesense.* the causative organism of sleeping sickness.

Scientific name	Common name	Mode of infection
Phlebotomus argentipes p. pepatasii and *p. sergenti*	Sandflies	Transmit the protozoan parasites, *Leishmania tropica, L. donovani* the causative organisms of Leishmaniasis (Kala-azar). Sandflies are responsible for oriental sore and sandfly fever also.
Sarcophaga spp.	Myiasis-producing flies	Maggots penetrate wounds and cause 'myiasis'
Siphunculina funicola	Eye-fly	Adults feed on secretion in the corner of eye. These are attracted to soars, cuts, ulcers etc. also.
Siphunculata		
Pediculus humanus corporis	Body louse	Cause considerable itching and scratching and transmit several diseases viz., epidemic typhus, trench fever and relapsing fever.
p. humanus capitis	Head louse	
Phthirus pubis.	Crab louse	
Siphonaptera		
Xanopsylla cheopis	Oriental rat fleas	Adults of both the sexes transmit plague bacteria *Pasteurella pestis*
Hemiptera		
Cimex hemipterus C. lectularius	Bed-bugs	Suck blood during night. Known not to transmit any disease but cause serious loss of vitality. Some are allergic to bed bug bites and develop rashes (Pruthi,1969).

TABLE 5.2: BENEFICIAL INSECTS
A. USEFUL MATERIALS FROM INSECTS/INSECT BODIES/INSECT BODY PARTS

Scientific Name	Common name	Usefulness
Hymenoptera		
Apis spp.	Honey bees	
Apis dorsata	The Rock bee	Honey is produced from the nectar
A. cerana indica	The Indian bee	collected by the worker bees. In India, now there are bee colonies
A. florea	The little bee	of all the four species of *Apis*.
A. mellifera	The Italian bee	Yield of *A. florea*, is poor hence it is uneconomical to domesticate it. *A. dorsata* is very ferocious and has not been domesticated inspite of its yield being higher than the other two Indians species. The Italian bee, *A. mellifera* was introduced in the Punjab and Himachal Pradesh in 1964 and its yield was found to be about four times than that of *A. cerana indica* (Atwal and Sharma,1970). The Italian bee hives are now in Pusa (Bihar) and possibly else where also in the country. Alcoholic extract from the bodies of honey bees is used in treating diseases like diptheria, scarlet fever, dropsy etc. (Frost, 1942). Bees wax is another useful product obtained from honey bees; used in the manufacture of many items including tooth pastes, shaving creams, face creams, lotions, cosmetics, boot polishes, paints, inks etc. (Fig. 66A, B)
Lepidoptera		
Bombyx mori (Linnaeus)	Mulberry silk worm	Reared on mulberry leaves, silk obtained is white or yellow. silk is the prized-fabric and is utilized for the preparation of parachutes, insulation coils for telephones, garments, sarees, shawls, etc (Fig. 67 A, B).
Samia cynthia *ricini* (Boisd)	Eri silk worm	Reared on castor leaves. Moths come out after cutting the cocoons. Cocoons cannot be reeled, have to be spun

FIG. 66A: Honey bee Queen surrounded by workers

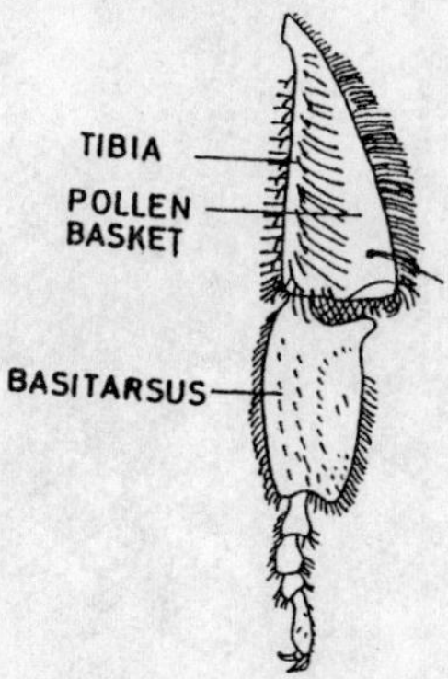

FIG. 66B: Leg of a honey bee worker showing the pollen-collecting apparatus (After Chapman, 1971)

FIG. 67A: Mulberry silkworm, *Bombyx mori*
(A) caterpillars feeding on mulberry leaves

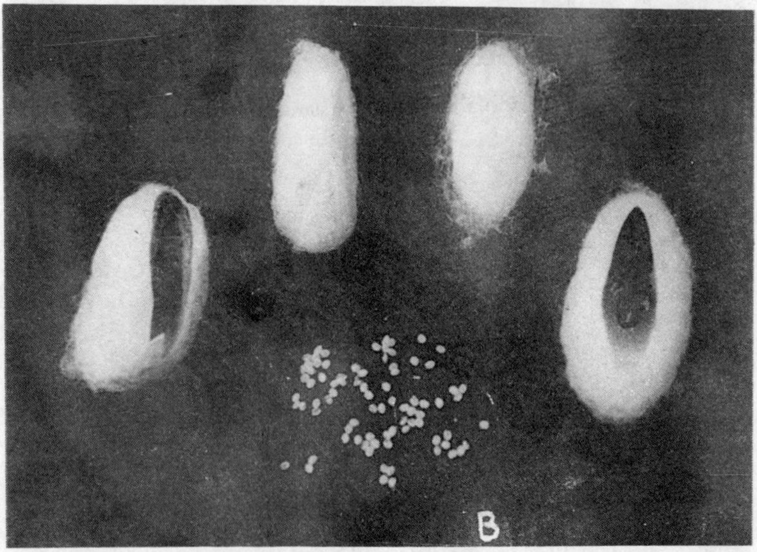

FIG. 67B: Eggs and cocoons of *Bombyx mori*

Scientific Name	Common name	Usefulness
		and woven like cotton.
Antheraea assamensis	Assamese silk worm	A semi-domesticated species, found mainly in Assam; yields muga silk.
A. yamamai	Japanese oak silk worm	Reared on large scale in Japan.
A. pernyi	Chinese silk worm	Commonly reared in China, silk obtained is pale buff coloured.
A. Mylitta	Tassar silk worm	Common host plants are *Shorea robusta* (Sal) and *Zizyphus mauritiana* (Ber). Moths do not pair in captivity and caterpillars cannot be domesticated. Cocoons are collected from the host trees.
Hemiptera		
Kerria lacca	Lac insect	Lac (Fig. 68) is the resinous secretion of the scale insect *kerria lacca.* Its main host plants are *schleichera oleosa* ("kusum") *Acacia catechu* ('khair' or 'kattha'), *Butea monosperma* ('palas' or 'dhak'), *Zizphus* sp. ('ber') and *Ficus* sp.

FIG. 68: Lac encrustation on Ficus twigs

Scientific Name	Common name	Usefulness
		Most of the lac production is from India. Lac is used for preparing gramophone records, electrical goods, toys, buttons, sealing wax, and paints and varnishes. However, for the manufacture of many of the items, these days, plastics is being used. Lac dye is obtained by extracting crushed 'Stick lac' in water and is still used in certain parts of India for dyeing wool and silk.
Coccus cacti	Scale insect of prickly pear and other cacti	Cochineal—a ''beautiful red pigment is obtained from the ground up bodies of *C. cacti*. The pigment is used to give beautiful colour to beverages, cakes, and medicines. Also used as cosmetics (Frost, 1942).
Diptera		
Cynips gallae-tinctoriae	Aleppo gall	Aleppo gall produced on several species of oakes in Eastern Europe and Western Asia is the chief source of ingredients for inks. This gall is also used to dye the hair black (Fros, 1942). Tannic acid obtained from galls is used in the preparation of Nicotine tannate which is used as insecticide.
	'Mad apple' gall	This gall of Asia Minor yields the dye Turkey red (Frost, 1942).
Coleoptera		
Lytta vesicatoria *Mylabris Cichorii*	Blister beetles	Cantharidine obtained from the bodies of beetles (Fig. 69) is used in the internal treatment of certain diseases of the urinogenital system. Cantharidine used in hair washes also but is considered harmful to hairs (Frost, 1942).

B. AS FOOD FOR MAN AND HIS ALLIES

Orthoptera		
Schistocerca gregaria	Locusts and grasshoppers	Locusts and grasshoppers are more commonly used as a part or human diet in many countries because of being comparatively

Scientific Name	Common name	Usefulness
Locusta migratoria and other species of locust and some species of grasshoppers		more commonly available. Locusts form a popular dish in Arabia, most of middle East countries, Baluchistan and Sind (Pakistan) and Western Rajasthan of India (Pruthi, 1969). They are dried, smoked/ salted and stored for long periods. Bheel and Meghwal tribes eat locusts and store in gunny bags The Greeks ground locusts and made flour out of it (Fig.9).
Isoptera		
Odontotermes sp.	Termites	Termites are eaten especially the queens by peoples of the tropical regions. Workers digging fields when come across termite queens just catch and eat (Fig. 25)
Atta cephalotes	Sauba ants	These ants are a part of breakfast of the natives of Amazan Valley (Frost,1942).
Corixa femorata		Eggs of this insect are fried and eaten and considered to be a delicacy in Mexico.
Ephydra hians	koo-tsabe fly	Shells of puparia of Koo-tsabe are removed and the deshelled pupae eaten.
Periplaneta sp. and *Gryllodes* sp.	Cockroaches and crickets	Eaten in some countries including Japan; even the live specimens are cut into pieces and eaten.

In addition to these, some other insects also are eaten by men e.g., full grown grubs of Goliath beetle measuring about 14 cm are eaten by Red Indians of U.S.A. Borers also are eaten.

Besides men, insects form diet/part of diet of certain animals and thus help man by reducing the number of insects.

Certain fishes like *Gambusia, Aplocheilus* and *Puntius* feed upon mosquito larvae, amphibians feed upon insects including pests of crops, reptiles like *Uromastix* feed voraciously upon locusts. Other reptiles e.g., *Chameleon* sp. and wall lizard feed upon insects which include crop pests and those affecting human and animal health. A large number of bird species like the Indian Myna, paradise fly catcher and cattle Ergot etc., feed upon insects affecting crops and human/animal health. Certain mammals like spiny Ant Eater of Australia and shrews are insectivorous. Thus it is evident that insects serve as food for many animals/birds i.e.,

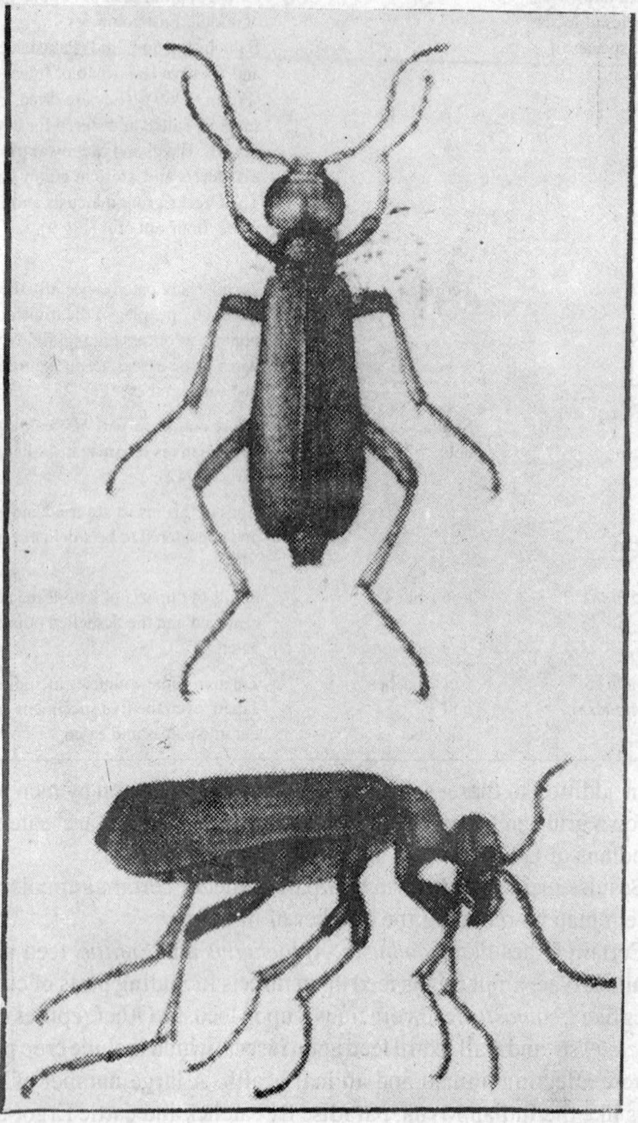

FIG. 69: Blister beetles

they are a part of diet of certain animals/birds which are friends of man.

Insects also serve as food of animals valuable to man. Many of our fishes live mainly on aquatic insects like blood worms, mayfly nymphs, and cadice worms. Many prized song birds and game birds feed to a great extent on insects. Insects form a part of diet of ducks, hens and cocks. Hogs feed and fatten upon white grubs (Frost, 1942).

C. AS MEDICINES

In addition to the use of alcoholic extract of honey bees as medicine in certain cases, the poison of honey bee sting is reported to help in "curing arthritis,rheumatism etc." (Pruthi,1969). The cocoons of the mulberry silkworm *(Bombyx mori)* are used as a tonic and its extract is reported to be used as a medicine in cases of leucorrhoea and chronic diarrhoea. Sharma (1990) has reported that Bheel tribe of Rajasthan eats silkworm larvae fried in mustard oil in cases of apetysis whereas larvae of the Tassar silk worm *(Antheraea mylitta)* fried in cow fat is used by the Bheels in cases of anaemia.

The cochineal insect *(Coccus cacti),* in addition to yielding dye has antiseptic properties and is used in cases of neuralgia and whooping cough. The blister beetle *(Mylabris* spp.) gives us cantharidine which is an internal stimulant and diuretic whereas the water beetles of families Dytiscidae and Hydrophilidae are reported to be anti-diuretic (Pruthi, 1969). Bheels of Rajasthan are reported to be using *Schistocerca gregaria* in cases of tuberculosis and bronchitis (Sharma, 1990). Slone (1991) has reported that in the encyclopedia of traditional Chinese medicines and in ancient Chinese records, in addition to the use of bee and silkworm products, hornets, wasp, mantids, flies, stink bugs, caterpillars, beetles, cicadas, mole crickets, silver fish, cookroaches, crickets, dragonflies and locust also are used as medicine. Chitin of insects e.g., of cicadas is also reported to have medicinal properties (Slone, 1991).

Certain species of cockroaches, bed bugs, blister beetles, predatory beetles, *culex* mosquito, ants, wasps, cochineal insects, honey bees and silkworms are used in the preparation of Homoeopathic medicines also (Boericke, 1991).

D. AS POLLINATORS

Pollination by insects can be considered to be a mutual association of insects and plants. Insects visit flowers to obtain nector and pollen

(Fig. 65) and while doing so, unconsciously pollinate the flowers. More than 1000 species of insects are known to visit a variety of plant species and pollinate the flowers which help in settling of seeds and subsequent fruit formation (Pruthi, 1969). Ornamentals like *Chrysanthemum,* orchids, *Yucca,* vegetables like peas, beans,tomatoes, cruciferous and cucurbitaceous vegetables etc., figs and *Ficus* sp. are pollinated by insects.

Insect pollinators belong to many orders namely Orthoptera, Diptera, Coleoptera, Thysanoptera, Lapidoptera, Hemiptera, Neuroptera and Hymenoptera. Some of the common insect pollinators are: Flower flies like *Eristalis* sp., *Syrphus* sp., *Platychirus* sp., *Rhingia* spp., flesh flies like *Calliphora* sp., *Lucilia* sp.,. *Sarcophaga* spp. However, bees are outstanding pollinators.

Names of some of the common insect pollinators and the crops they pollinate are given below:

Scientific Name	Common name	Names of plants whose flowers are pollinated
Apis spp.	Honey bees	Cruciferous crops, like mustard, cole crops, radish, turnip, etc., cucurbitaceous crops like pumpkin, bottle gourd, *Luffa,* water melons, musk melons; leguminsous crops like pulses, beans; oil seed crops like mustard, sunflowers, linseed, sesamum; fruit crops like apple, pear, litchi, orange, grapes etc. Bees have been reported to forage upto 800-900 meters. Experiments at Rajendra Agricultural University, Pusa (Bihar) have indicated that due to crops-pollination by honey bees, eliminating all other insects, there was an average increase of 178 per cent in the yield of onion, 131 per cent in mustard, 675 per cent in sunflower, 511 per cent in safflower, 232 per cent in linseed, 500 per cent in sesamum and 35-90 per cent in orange (Sinha and Singh, 1988).
Megachile nana Bingham *M. flavipes* Spinola	Megachilid bees	*M. nana* was one and a half times better than *M. flavipes*

Scientific Name	Common name	Names of plants whose flowers are pollinated
M. femorata Smith		in pollinating the flowers of alfalfa *(Medicago sativa)* (Kapil and Jain, 1980).
Bombus sp.	Bumble bees	Clovers (Imms, 1965)
Blastophaga spp.	Fig wasp	Figs (Frost, 1942; Pruthi, 1969)
Tetraopes tetropohthalmus	Milk weed beetle	Milk weed *(Asclepias syriaca)* (Frost,1942)
Tegeticula (Pronuba) sp.	Yucca moth	Yucca flowers are pollinated by the female moth whose proboscis, unlike other Lepidoptera, is not long and curled, but is short and spiny, modified to collect pollen (Frost, 1942).
	E. AS PRODUCERS OF LIGHT	
Lamprophorus sp. *Photinus marginellus*		Two families of Coleoptera, Lampyridae and Elateridae have species which emit light.
P. scintillans *Photuris pennsy-lvanica* *Microphotus augustus* light	Glow worms and fire-flies	Males are winged and are known as fire flies and the wingless females and grubs as glow worms. Genus *Lamprophorus* is commonly found in India. Rays emitted by fireflies and glow worms are 92-100 per cent light and only 0 to 8 per cent are heat rays hence the produced by these beetles is cold light. Light produced by other sources has 3 to 25 per cent light rays and the rest are heat rays hence these lights are not cold lights (Frost, 1942).

Note: Bees had to be imported into Australia to make the cultivation of red clover possible there.

F. AS BIOLOGICAL CONTROL AGENTS OF INSECT PESTS

There are a number of insects which help man by feeding upon/killing insect pests. These are known as parasites (*Trichogramma* spp., *Bracon brevicornis, Paraicorola nephantidis* etc.) and predators (*Mantis* sp., Fig. 70) *Syrphus, Coccinella* (Fig. 71) *Cicindella* etc. (Fig. 72) This beneficial aspect of insects has been dealt with in detail in chapter 8.

G. AS BIOLOGICAL CONTROL AGENTS OF WEEDS

The prickly pear *Opuntia* spp. has become an aggressive weed in many countries. Caterpillars of the moth *Cactoblaslis cactorum* feed by boring into *Opuntia*, thus allow easy penetration of bacteria and fungi which destroy the plant completely. In Australia, USA, Mexico and Argentina several million hectares of land have been cleared of this weed and brought under cultivation (Arora and Sharma, 1990). In Hawaii C. *cactorum* and the mealy bug *Dactylopius opuntiae* have proved effective for controlling prickly pear (Pruthi, 1969).

According to David and Muthukrishnan (1953) prickly pear *Opuntia vulgaris* and *O. elatior* were introduced into India about 1787 for use in cultures of the commercial dyes producing cochineal. *O. vulgaris* did not become a serious pest because of the unnoticed introduction of the cochineal insect *Dactylopius indicus. Dactylopius tomentosus* was introduced in India in 1926 from Ceylon (Sri Lanka). This insect established and gave good control of *Opuntia* spp. which has been controlled in over 100,000 acres in South India.

Gorse *(Ulex europeus),* a yellow-coloured perennial introduced as an ornamental became a menace to reforestation in California, Australia, New Zealand, Hawaii, etc. The gorse seed weevil *Aion ulicis* was introduced in these countries from England. There are reports of its seed potential having been reduced upto 98 percent in New Zealand (Thomas, 1970).

The goat weed, *Hypericum perforatum* was controlled in California by introducing the goat weed beetles, *Chyroline hyperici* and *C. guadrigemina* from Australia in 1945 in California and some other places (Arora and Sharma, 1990).

Another important weed, the Alligator weed, *Alternanthera philoxeroides* has been controlled effectively in Florida by introducing the flea beetle *Agasicles hygrophyla* (Arora and Sharma, 1990)

Lantana camara is another important weed which has spread mainly in northern India. A lace bug *Teleonemia scrupulosa* was introduced but during specifically tests it was found to attack teak also (Gardner, 1944). However, a lantana seed fly *Ophiomyia (Agromyza) lantanae* introduced from Hawaii was considered to be an efficient natural enemy of lantana, but according to Pruthi (1969) it did not establish in India. Lohumi (1980) has, however,controlled *Lantana camara* in Almorah District of Uttar Pradesh with the help of the lace bugs *T. scrupulosa.* According to him, it does not feed on teak leaves. Arora and Sharma (1990) also consider it to be most effective insect to

FIG. 70: A preying *Mantis* predating on a desert locust hopper

FIG. 71: Lady Bird beetles—effective predators of aphids

control Lantana.

Water hycinth, *Eichhornia crassipes,* an aquatic weed has been controlled in India, to a great extent by the weevils, *Neochentina eichorniae* and *N. bruchi,* introduced from Latin America. These weevils have given about 95 per cent control in areas in Bangalore (Jayanath and Nagarakatti, 1987). Another weevil, *Cyrtobagous salvininae* introduced from Australia is giving 99 percent control of another weed, the water fern, *Salvina molesta* in Kerala (Joy *et al.* 1985).

Field trials are being conducted in nine states in India for the control of the congress weed, *Parthenium hysterophorus* with a chrysomelid beetle *Zygogramma bicoloralia* introduced from Mexico. (Jayanth, 1987).

H. AS SCAVENGERS

Insects which feed upon waste material are known as scavengers. The waste material may be dead animal or plant material or it may be excrement of animals. Scavengers help us in removing the waste material, and thus contribute towards reducing pollution. Frost (1942) has divided scavenger (Saprophagous) insects into three groups:

(a) Phytosaprophagous
(b) Zoosaprophagous and
(c) Scatophagous

(a) *Phytosaprophagous* are those which feed upon dead/decaying vegetable material. Thysanura, Collembola feed largely upon decomposing vegetable material, larvae of many Diptera feed upon decayed plant material. Insects attacking decaying/ dead wood are certain beetles e.g. Passalus, Cupes Sacarabaeidae and Elateridae, certain Diptera e.g., Tipulidae, Rhagionidae and Asilidae. Termites and powder pest beetles attack even dry and cured wood.

(b) *Zoosaprophagous* are those which feed upon dead animal material. Larvae of the water beetles of Hydrophildae feed to some extent upon insects that fall upon water *Volucella* feeds upon dead insects in the nests of social insects. Larvae of *Galleria* sp. feed upon nitrogenous material in the comb, consisting of molted skin and other materials associated with bees (Frost, 1942).

(c) *Scatophagous:* Insects belonging to this group feed upon

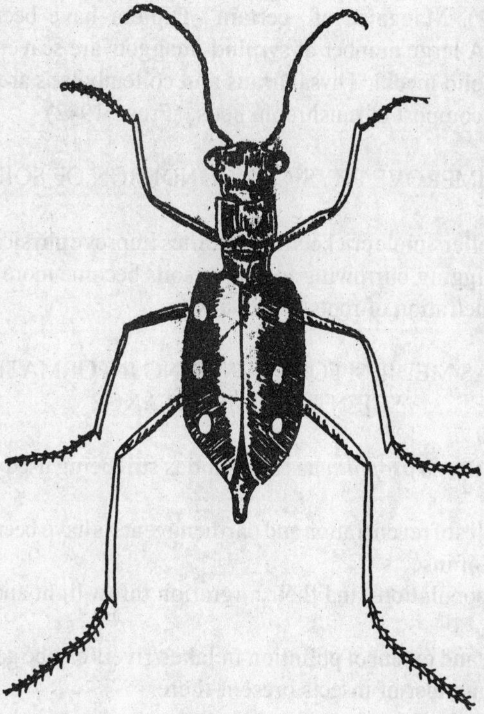

FIG. 72: *Cicindella* sp., an effective predator of gundhy bug *Leptocorisa* sp. (After Pruthi, 1969)

waste material. The young of the sticktight flea, *Echidnophaga gallinacea* feed upon the faecal matter of the adults.Diptera and certain scarabeid beetles have been found in and on dung of animals. The dung of placentals is rolled into balls by the dung roller beetles and buried underneath the soil.Dung rollers lay eggs in dung and the hatching grubs feed upon dung.

The dung rollers of Australia are adapted to the hard pellet, like droppings of endemic mesupials (e.g., Kangaroos). Thus the dung of the placentals like cows, buffaloes, horses etc. introduced in Australia from other countries remain uncleaned and their non-clearance had been creating many problems. Dr. Bornemissza of CSIRO Australia was on a visit to India and South Africa in 1972 to collect the dung rollers and despatch to his country for the clearing the dung of the placentals (Menon,1972). Maggots of certain Diptera have been found in excrements. A large number of syrphid maggots are scavengers living in liquid or solid media. Thysanurans and collembolans are commonly found in the compost of mushroom beds. (Frost, 1942).

I. IMPROVE PHYSICAL CONDITION OF SOIL

Ants, dung rollers,mole crickets and termites improve physical condition of soils, by digging/burrowing in soil as soils become more porous and make the penetration of roots in soil easier.

J. ACT AS MEDIUM FOR PROVIDING INFORMATION OF SCIENTIFIC IMPORTANCE

The pomace fly, *Drosophila* has been and is still being used in genetical studies.

Principles of regeneration and parthenogenesis have been discovered by the study of insects.

Insect populations and their migration throw light and ecological conditions.

Degree and extent of pollution in lakes/rivers can be gauged by the species and number of insects present there.

The length of time an animal has been dead can be determined by the insect species present on/in dead bodies.

K. AESTHETIC AND ENTERTAINING IMPORTANCE

A number of insects produce noises of different types in various ways—by rubbing wings against legs, by rubbing wings against wings, by vibrating their wings rapidly and in many other ways. To some, these noises appear to be pleasant and they give it the name of music while others call it unpleasant noise.

The famous English poet Shelley was happy to hear the 'song' of grasshoppers and calls it 'one joyous thing in a world of sorrowing' (Frost, 1942). Charles Dickens was inspired by the song of crickets (Frost, 1942).Nicolas Rimsky—Korsakov wrote the poem 'Flight of the Bumble Bee''.

Joseph Strauss got inspiration from the zig-zag flight and swift movements of dragonfly and expressed it in the 'Dragonfly''.

Edvard Crieg expressed his inspiration, obtained from the butterfly in "Papillon"

"The Beetles' Dance", "Minuet of a Fly" and "Song of the Flea" are other instances of poets getting inspiration from insects.

In Japan, cicadas and crickets are caged like birds in small cages. These are also sold in markets. Every year, during the course of the celebration of 'The festival of singing insects' for a week all the caged insects are released to give them freedom for some time.

Beautiful colours, shapes and patterns of wings of butterflies act as models for artists, designers and decorators. Jewellery and pictures made from pieces of wings are durable and the colours do not fade. Certain tribes in Ecuador make earings from the elytra of bupestrid beetles. Egyptians selected scarabid beetle as a symbol of God Khepera. In Greece insects like bees were inscribed on coins. (Frost, 1942)

In Japan insects e.g., Cicadas, crickets and dragonflies are seen on the carvings of ivory, rock crystals, and wood. Red Indians of America used insects to decorate pottery and baskets etc. Insects are seen on printer's mark, book marks, water marks and postage stamps (Frost, 1942). In the modern world beautiful insects especially butterflies in their natural colour patterns find place in Greeting cards and Wedding cards etc.

6

INSECT PESTS AND THEIR ORIGIN*

A. DEFINITION OF A PEST

A pest has been defined by Conway (1976) as a living organism which causes damage or illness to man or his possessions. Pests compete with man for food, fibre and shelter, transmit pathogens, threaten human health, comfort or welfare (Flint and van den Bosch, 1981).

Pests include insects, mites, ticks, nematodes, molluscs, crustaceans, birds, rodents, fungi, bacteria, viruses and weeds.

In the natural habitat of an organism, it may not be in conflict with man but its status may change especially when man practises monoculture, growing a single crop in a large area.

Weeds compete with agricultural crops for nutrients, water and light especially in the early stages of growth and the quality or quantity or both, of the crop produce are affected. 'Key pests' are those which cause major damage to crops every season unless checked, whereas pests which cause damage at irregular intervals are known as 'occasional pests' (Mathews, 1984).

For determining the status of a pest, it is the number and frequency with which it occurs rather than nature of damage which matters.

Let us illustrate it with a few common examples:

(a) The red cotton bug, *Dysdercus koenegii* is a pest of cotton and lady's finger; neither its number nor its frequency is much and it causes some damage by sucking the sap from stems and fruits etc., hence it is not considered as an important pest.

---

* Based mostly on Late Dr. S. Pradhan's thoughts.

(b) The gundhy bug *Leptocorisa* spp. is a pest of paddy and causes damage by sucking the sap from the grains but its number is so large that its sucking the sap from the milky stage of paddy may result in almost complete crop failure. Hence this is considered to be a key pest and due attention is paid to its control in time.

(c) Aphids also, like *Dysdercus* and *Leptocorisa* cause damage by sucking the sap from plant parts e.g., leaves, stems, inflorescences and green pods etc., sometimes even to the extent of necessitating resowing. This pest lasts for several months. In this case also control measures have to be undertaken in time.

(d) Scale insects also cause damage by sucking only. Their population is very large, they stick to the plant almost collectively and cause such a severe damage in certain cases, that the twigs on which these are aggregated or the plant as a whole ultimately dies.

All the four insects mentioned above belong to the same taxonomic group — order Hemiptera and their nature of damage is the same (cause damage by sucking plant sap of the host plant) yet their extent of damage varies so much that the red cotton bug is categorised as a minor pest, while the other three as major pests.

B. BIOTIC POTENTIAL AND ENVIRONMENTAL RESISTANCE

As already stated earlier in Chapter 4, biotic potential of most of the insects is so high that the progeny of a single moth can cover the entire dry surface of earth to a depth of 24.32 metres in one year if each moth lays even 200 eggs (many moths lay much more than 200) and each one completes its life cycle in one month only. To illustrate this with an example, *Earias vittella* lays about 200 eggs and the life cycle is completed in about a month. Since one pair lays 200 eggs, the progeny consists of 100 males and 100 females; then in one month 100 x 200 =20,000 (10,000 males + 10,000 females). Thus in one year the population could be 2,000,000,000,000,000,000,000,000. This could have happened if the entire progeny of all the progenies would have survived. However, since the progeny meets with powerful environmental resistance majority of them die and are never able to cover any part of earth to any depth.

Let us see how slight variation in mortality makes a big change in pest population.

1. No. of eggs laid by an insect 1000
2. No. of progeny not reaching adult stage (Total mortality) 998
3. Thus the number reaching adult stage 1000–998=only 2
4. Percentage of mortality 99.8

1. No. of eggs laid by an insect = 1000
2. No. of progeny not reaching adult stage (total mortality) = 996
3. Thus the number reaching adult stage = only 4.
4. Percentage of mortality thus = 99.6

Thus, with a very slight decrease in the mortality of 0.2 per cent (from 99.8 to 99.6), the population must double up from 2 to 4.

The high biotic potential of insects has to face a more or less equally high environmental resistance, which insects devise methods to face boldly, on account of which atleast a very small percentage is able to survive and even this insignificant survival finally leads to their numerical domination. Population of an insect depends upon its biotic potential as well as upon the environmental resistance it has to meet. Thus,

$$\text{Population of an insect} = \frac{\text{Biotic potential}}{\text{Environmental resistance}}$$

The environmental resistance is of two types :

(i) Biotic
(ii) Abiotic (Ecological).

(i) Biotic Resistance

Each pest has certain natural enemies in the form of predators, parasites and pathogens which play their role in reducing the population depending upon their own biotic potential and the extent to which they are effective.

To escape from the biotic threat, insects have devised ways and means to dodge their natural enemies. Most insects have developed these devices; but all have not succeeded in dodging their natural enemies to the same extent. The natural enemies do take a toll of insect pests depending upon several factors.

(ii) Abiotic (Ecological) Resistance

The main ecological factors are the extremes of heat and cold i.e. extremely high and extremely low temperatures. Since they are cold-blooded creatures they cannot regulate their body temperatures hence they enter into a state of inactivity which is technically known as aestivation or hibernation. Although these devices of entering into the states of inactivity are no guarantee for the resumption of activity on the return of favourable conditions, yet it is a safeguard for the continuance of the species and many do resume their activity on the return of favourable conditions. Normally the number of insects resuming activity depends upon the duration of hot or cold spell, the longer the spell the lesser the number resuming activity. In certain social insects only the queen undergoes hibernations, others of the colony die and on the resumption of activity the queen produces individuals as required and then the new colony starts functioning.

Other factors are humidity and availability of food. Problem of humidity is solved by hibernation/aestivation whereas the food problem is solved by changing the food habit, shifting to alternate host plants or by migration. Some of the examples of migration are locusts, *Agrotis, Papilio* and aphids.

C. ORIGIN OF INSECTS WITH ORIGIN OF PLANTS

Insects originated with the origin of plants. It is a fact of history that originally, all life was aquatic and as land was formed, plants and insects both migrated on land. Those which could adopt to terrestrial life survived.

Plants dispersed on land and produced long stems, flowers and fruits. To feed upon these , insects had to leap and their legs got modified to reach the plants and finally developed wings to reach the plants of their choice. Thus, acquisition of wings and consequent power of flight was the most important modification in the evolution of insects from aquatic to terrestrial and aerial existence. Thus the origin of insects coincided with the origin of plants. With the modification in their morphology, changes in the morphology of insects took place to suit the change in the environmental conditions. Plants and insects have ever since been fighting for their survival. The long history of insects' struggle has resulted in several morphological changes in insects. However a natural balance was maintained as a result of which plants and insects both co-existed without any apparent disturbance to each other. So far man had not entered the scene.

D. ORIGIN OF PESTS WITH ORIGIN OF AGRICULTURE

When man entered the scene much after insects and plants had already existed, he upset nature's horticulture and insectary by planning according to his needs. He cut jungles after jungles for cultivating plants of his choice thus depriving numerous insects of their food and shelter. Till then there was mixed plant growth but man started the cultivation of plants of his choice and one-type plant crop system (monoculture) in which plants of only one type were grown.

A clear case of migration and change of food habits as a result of cleaning forests was observed recently in Baharaich District, Uttar Pradesh (India). A few years back the State's Farm Corporation of India cleared 600 hectares (1,500 acres) of forests in Baharaich District to establish a farm. In the cleaned area mustard was sown but the crop was attacked by the hairy caterpillar, *Achaea janata* which was earlier feeding on some plants in the forest, but got dislodged and the insect had to shift to mustard for survival. The attack was so severe that the mustard crop, in the early stages itself was ruined completely. Chemical control measures failed to save the crop even though the pest got killed. The crop was resown but it got reinfested due to further migration; ultimately the farm people had to change their pattern of cultivation and shifted to growing of underground crop like turmeric.

Insects which were displaced had only three choices open to them: (a) to starve to death, (b) to migrate to other plant or (c) to change their food habits. Those which could shift to new plants grown by man i.e. to his agriculture and adjusted to this type of food, cultured by man found it convenient and fed and bred well in time and thus got rehabilitated. Such insects found it very congenial and the risk of searching the food for their survival was covered up and their number swelled. Since man's efforts to grow plants were being nullified by insects, man started devising ways and means to keep the insects away from their field or to kill them whatever could be feasible. Such insects which survived on man's agriculture were thus labelled as pests and their food plants were labelled as their host plants. Thus the beginning of war between insect and man started.

E. INTENSIFICATION OF AGRICULTURE INTENSIFIED PEST PROBLEMS

With development of medical science, child mortality in particular and mortality in general, decreased leading to increase in human population and resulting in more demand for food. As demand for food went on increasing man started bringing more land under cultivation. As man's food requirement went on increasing with the ever increasing human population, man planned to meet this additional requirement in the follwing ways :

(a) By growing more than one crop in the same field in many cases the same crop in succession, in others other crops, leading to taking 2-4 crops (from field in which there used to be only one crop) leading to perpetuation of pest population (Table 8.1)

(b) By growing short duration varieties to get quick yield in a short period leaving the land free for growing other crops in the same field.

(c) By increased use of fertilizers to recoupe soil fertility. All these measures adopted by man to intensify agriculture for getting increased production were favourable even for insect growth hence it led to intensification of pest problems also.

As a result of breeder's experiments without coordination of Entomologists, a number of high-yielding varieties in most of the important food crops as well as commercial crops were produced. Thus creation of high-yielding varieties led to the 'narrowing of germplasm' because most of the indigenous germplasm is perishing as yields of the new varieties/hybrids are so significantly high that all have rushed to cultivate these, leaving the indigenous ones to perish—the varieties which had existed since long and had stood the pest attacks boldly. Now there is continuity of the same crop and same variety e.g., the crosses of Mexican wheat are grown continuously from Assam to Afghanistan and it has become still more congenial for the pest population to continue to exist and thrive. Likewise certain new rice varieties are being grown all over from Cuttack to Manila. The low-yielding varieties which had stood since long even before the pesticidal era might have contained the germplasm of resistance against insect pests.

The newer varieties are generally highly susceptible to pests and cannot exist without pesticidal umbrella. Had there been adequate coordination between Plant Breeder and Entomologist, this unhappy situation would not have arisen and the breeder might have been able to incorporate the germplasm tolerant to atleast some major pests. Thus many of the potential pests which had remained suppressed for long, started raising their heads with the availability of more food of their choice continually from one end to the other. Thus *Heliothis* which till recently was known to attack only gram has started attacking other crops also e.g., the American cottons in Maharashtra, Gujarat, Karnataka, Tamilnadu and Andhra Pradesh; it has also now been reported to be becoming pest of wheat in Delhi, Himachal Pradesh and Punjab. This pest was reported on wheat in 1919 but it had the importance of being only a new host record. *Pyrilla* which was occurring after gaps of few years on sugarcane is now standing almost on a permanent footing not only on sugarcane but also on some other graminaceous crops like sorghum, maize, wheat and paddy and a continuity of the pest is being maintained.

The paddy leaf hopper, *Nilaparvatha lugens* has been recently reported in epidemic form in certain Southern districts of Kerala and has been reported to be in more serious form on "improved varieties" of paddy receiving heavy doses of nitrogenous fertilizers. It has been reported from Delhi region also on paddy.

Earlier when a crop was harvested, the pests hibernated, migrated or died. During migration as well as during hibernation/aestivation many perished. Now growing of one crop after another and in a number of cases the same crop, say paddy after paddy in succession, has, while giving the cultivator the opportunity to take 2-3 crops instead of only one from a field, further perpetuated the continuity of pests which would have otherwise, atleast dwindled if not perished completely. They continued to get congenial environment and there were many more generations of the pest than otherwise would have been.

Growing of short duration varieties is also responsible for accentuating pest problems, in certain cases. Short duration varieties of pigeon pea namely T-21, Prabhat and S-5 (Ageti) which mature during October-December were observed to suffer badly in Delhi, U. P., and Bihar due to *Maruca testulalis* Geyer which is already a serious pest of lelgume crops in peninsular states (Saxena, 1974). In long duration varieties of 'arhar', pod formation takes place during winter months and by March-April, they are harvested hence there is no scope for the attack of this pest on 'arhar' of long duration varieties.

Increased use of fertilizers to recoup soil fertility, while increased use of fertilizers has resulted in the better and healthier crops, it has increased pest problems also. Due to the increased use of nitrogenous fertilizers plants develop much better succulent vegetative parts and attract more insects, i.e., it has been observed that *Pyrilla perpusilla* infestation is much more severe in those sugarcane fields which have broader and succulent leaves and rich in nitrogen content. However there may be some exceptions, e.g., the white fly of sugarcane *Alearolobus barodensis* is more severe in the crops which are low-lying and the soil also deficient in nitrogen content.

Thus the way intensification of Agriculture has taken place, has led to the intensification of pest problems also. This statement can be further amplified when we peep into the crop yields of various crops during the last over 80 years, i.e., from the time the departments of agriculture were established in different states of the country. Due to the negligence of pest problems, there has been a prolonged decline in the crop yields per hectare which has not been possible to make up in most of the crops even now. In the graph plotted as year and yield (Pradhan, 1973), we find that crop yields/hectare continued to decline in all the crops till about 1950, after which it has risen in some crops. It is significant that many of the chemical pesticides belonging to practically all groups namely chlorinated hydrocarbons (DDT, BHC, aldrin, dieldrin, endrin), carbamates (carbaryl), organophosphates (parathion, malathion) and systemic insecticides (pestox, schradon) were introduced during 1940-1950 and started extending a helping hand to the plant Breeders and Agronomists.

The data also show that intensification of Agriculture did not result in increase in crop yield as it led to simultaneous intensification of pest problems. While the breeders, agronomists and extension workers were all along emphasising the use of various inputs like good quality seeds, manures, fertilizers, etc., they completely neglected the application of Plant Protection measures. Even where Plant Protection Technology was practised it was based more on analogy than on actual experimental data and in many cases by persons not trained in Pest Control Technology. Thus, whatever little pest control efforts were made, were not directed/ guided pest controls. Continued decrease of yield, however, for half-a-century focussed the attention of agriculturists on it, resulting in pest control revolution 22 years later.

Inspite of man's superiority over insects and for that matter over all the living creatures, because of the superiority of brain and intelligence, insects continue to dominate the scene. The war waged by man against pests is technically known as control. This is the only correct term as man can, at best, exercise a check on insect population and reduce its number to such a low level that it loses its status as pest and the damage to crop may be reduced to a negligible level.

7

BASIC PRINCIPLES OF INSECT PEST MANAGEMENT

The main objective of undertaking control measures is to maximize the yield and to reduce the losses caused by pests. The yield of a crop depends upon:

Yield

Production	Protection
Increasing production means making all possible efforts to get better yield by giving adequate inputs of fertilizers, irrigation etc.	Protection aims at making the life of pest hard and to reduce its population to such a low level as to dethrone it from the status of pest, taking care that the cost involved in the input does not exceed the cost of output.

Thus :

Yield potential = production potential—destruction potential due to pests.

Thus, it is evident that the basic principles involved in pest control are governed by the laws of economics according to which output—input = gain. Under these circumstances only, control measures can be practicable. Agriculture is an industry; the pros and cons of control must have definite relations with the economics of investment and production.

Losses can be of two types : felt losses and actual losses, the latter being more spectacular. Thus, economics of the production of any crop must be thought out carefully before undertaking any pest control operations.

A number of experiments have been carried out on the response of crops to various inputs like better variety, better seed, fertilizer application, fungicidal and insecticidal treatments. The data provided by these experiments have revealed that increases in yield due to pest control alone, are much higher than any other input used alone. Thus, in the tropical countries, protection research is more needed than production research as insects are much more abundant in the tropical than in the cooler countries.

Thus, there are three basic principles on which the success of pest control depends : prevention, cooperation and economics; if any of these is neglected or ignored success is bound to be cut short.

A. PREVENTION

The saying "Prevention is better than cure" is well known and holds good even in relation to pest problems. Though at the outset, it may not appear to be very promising, but the end result is very spectacular. Today we are in the age of chemical control and on account of the continual production of newer pesticides frequently, we do not think of any method other than the chemical. But if we look back to the old farm practices which are given the name of cultural control, these are nothing but just measures which prevent the outbreak of pests. Thus, clean cultivation, keeping the crops and their neighbourhood including bunds free from weeds, manipulation of the sowing time i.e. shifting the sowing time, ploughing the field to expose the pests hiding therein, to birds and other predators and shifting the position of egg pods to prevent hatching e.g., in case of 'Phadka' grasshopper, crop rotation to deprive the pests hibernating in the field and thus, stop access to their food plants; pruning,thinning and disposal of the infested material e.g.; pruning the brinjal plants infested with shoot borers, followed by burning the infested pruned shoots and plucked fruits, reduces the brinjal borer menace and also results in better branching leading to more fruit yield; use of resistant varieties e.g., varieties Co 205 and Co 285 of sugarcane are generally resistant to borers but susceptible to Gurdaspur borer, *Bissetia steniellus.* A resistant variety may be resistant in relation to one and susceptible to other or resistant in one region but susceptible in another region,and manure or fertilizer manipulation e.g., the sugarcane white fly attack can be reduced to a great extent by giving more nitrogenous fertilizers. Moreover preventive measures are cheaper and in some cases practically no cost is involved.

B. CO-OPERATION

Co-operation amongst the cultivators is important especially in the present era when we are growing continuously and in many cases the same crop of the same variety from one end of the country to the other and in certain cases even beyond the frontiers of the country. Co-operation is of vital importance in the case of migrating and long range flying insects also e.g., Locust when forms swarms, leaves its original place of swarm formation and moves to other localities; in some cases it settles on one crop, finishes and flies off and settles on another crop may be in a different state of a country or even in a different country that is why there is a network of locust control organization all over the country/countries liable to invasion, so that if a swarm has settled in one region, the neighbouring state/country may be informed to be careful. Food and Agriculture Organization (FAO) co-ordinates the activities of the main locust centres of different countries.

Another very common example of co-operation is the control of pink bollworm. It is a serious pest of cotton especially in Maharashtra and Gujarat—the two major cotton growing states. Cultivators are advised to dry the cotton seeds in the open sun to kill the hiding pink bollworm larvae. If by any chance even one cultivator does not practise the sun-drying process, the menace would spread to other fields also. That is why the Government has passed a legislation for compulsory sun-drying of cotton seeds prior to sowing. Co-operation in case of other crops also reduces the cost involved in undertaking control measures as it is a great check on the spread of pests. If all the cultivators of adjacent/neighbouring fields undertake control measures simultaneously the pest would be destroyed in a very short time as it would not get shelter even on short distance migrations. It has become all the more necessary under the changed circumstances in this era of high yielding varieties when we find huge tracts of land, crop or crops having similar /identical pest complexes. Thus, to exercise an effective check on pest population, to obtain successful pest control and to reduce or eliminate outbreak of pests and pest epidemics, co-operation is an important principle of pest control.

C. ECONOMICS

Agriculture like any other industry is based on economics and unless the total cost of the input involved in production is less than the output, it cannot run on sound footing. Agricultural production depends on land, labour, capital and organization. The control measures have to be so organised that the capital invested in control must result in increased production; then only can it be said to be economical and then only can it be said to be feasible. The economics of pest control is governed by the following factors :

Nature of the Commodity and its Market Value

If the crop attacked happens to be leafy vegetable like spinach and the insects attacking it happen to be biting and chewing type, the sale price is very much reduced. Similarly apples are attacked by scale insects resulting in numerous spots on the fruits, on account of which the sale price is very much reduced, although the nutritional value is not reduced proportionately. Thus, it is the market value and not the nutritional value which decides for undertaking or not undertaking control measure against a certain pest. Also, if the commodity is very cheap, the cultivator may not agree to undertake control measures. Thus, during summer season the cultivator may readily agree to undertake control measures against spinach pests but the same cultivator may not agree for it during winter season when the rate/kg. is very much lower than that during the summer months. Moreover, if the cultivator undertakes control measures during the summers, his winter crop is likely to be almost pest free; if the cultivator is made to understand this economics, he is bound to agree readily.

Type of Crop and Yield Potential

There are still a number of cultivators who do not invest much in manures, seeds, fertilizers and pesticides, and continue to practice their agriculture in the old traditional fashion. Such cultivators are either lazy or their wants are so limited that they remain contented with whatever yield they get or they are too poor to invest anything except the seeds or they are solely dependent on fate. On the other hand, there are other cultivators who are progressive and procure good quality high-yielding variety seeds, invest in manures, fertilizers and pesticides. Thus, if the crop is an improved variety and the yield potential is high, it would be profitable (economical) to undertake control measures.

Intensity of Infestation

If the intensity of infestation is low and the damage expected is negligible, it can be ignored; however, if the infestation is so high as to lower the yield substantially, it would be economical to undertake control measures. In the case of aphids, generally the intensity of infestation is so high that it may lead to almost complete crop failure. In such cases timely control operations have to be organised so that the crop may be saved. If possible, in case of aphids, the crop may be treated before the aphids start infesting inflorescences. Similarly when gundhy bugs appear in the field no time should be lost in undertaking control measures.

Nature of Damage

If the nature of damage is such that it can be tolerated e.g. mere surface feeding etc., one may afford to do without adopting any control measure. However in the case of vectors, which, in addition to causing damage, also spread the disease, timely control is indispensable e.g., the white fly of sugarcane and the paddy jassids.

Long-term and Advance Planning

If we know which pest is likely to appear when, in certain cases we do know, we can plan in advance to adopt timely control measure. Timely and well-planned control measures are both effective and economical. In some cases initially a certain amount of investment may have to be done e.g., in importing exotic predators/parasites or in laboratory breeding of exotic/indigenous predators/parasites for releasing them in the field with the object of establishing them in the natural habitat of the pest. Such investment would, in the long run, prove profitable and economical as it would eliminate or atleast reduce the expenditure involved in regular pesticidal treatments.

8

INSECT PEST MANAGEMENT

A pest may be defined as an organism which competes with mankind for limited resources, or threatens man's health or comfort and possessions. It has been estimated that ten thousand insect species are considered economic pests. This number is, thus, only one per cent or one in each 100 species (Wilson *et al.* 1984)

Pest management refers to actions which are primarily aimed at harming living organisms that are harmful to man and his resources. The management of pests is what is commonly meant by the term pest control. It is in fact an intelligent management of pests (Beirne, 1967). The pest management may involve simple manipulation such as spraying of roses with pesticides. Thus, the term refers to any practice or combination of practices designed to manipulate pest or potential pest population and to diminish pest injury or render pests harmless (Flint and van den Bosch, 1981)

The importance of a pest may be related to the value of damage it causes. The damage may be quantifiable e.g., reduction in yield or number of animals infected with parasites, though qualitative assessments are often necessary. The lower grade of produce or simply delay in harvesting can have a considerable impact on the price paid particularly in case of many vegetable crops that undergo price fluctuations.

Economic Injury Level (EIL) has been defined by Headley (1972a,b) as the pest population that produces incremental damage equal to the cost of preventing damage i.e., thc lowest pest population which will cause economic damage.

Economic Threshold (ET) is the population density at which control measures should be determined to prevent an increasing pest population reaching the EIL. The ET is lower than EIL so as to give sufficient time to put control measures into operation and have them take effect.

Our approach to insect pest management should be such by which the life of pest can be made hard leading to reduction of population to a very low or insignificant level with minimum amount of expenditure and yet have little or no side effects.

Insect pests can be managed in many ways which may broadly be classified into (a) natural control and (b) applied control.

Natural Control

- Climate
- Topography
- Naturally occurring natural enemies

Applied Control

- Cultural
- Mechanical
- Physical
- Chemical
- Biological
- Legislative
- Integrated

A. NATURAL CONTROL

Natural forces, which include both biotic and abiotic factors of the environment are very powerful and normally maintain the pest population within definable upper and lower limits. However, when these are disturbed, their functioning gets upset on account of which they are unable to keep a check on pest population; this naturally leads to increase in pest population. Various factors of natural control are as follows:

(a) Change of Climate/Weather

Climate/weather is perhaps the most important factor. It may be said that the insect life of a place is dependent directly or indirectly on temperature, soil and amount of moisture. Populaticn of several insect species is considerably reduced because of change in weather. For example, although mosquitoes occur throughout the year but their number is far less during extreme winter or summer. Hot and very dry climate is generally unfavourable to majority of insects. A warm climate with high humidity is unfavourable to many insects; such conditions favour the development of diseases. Wind velocity is also of great importance. Many species of insects are not able to reach the plant because of high wind velocity and any attempt to move may,sometimes prove fatal.

(b) Change of Topography

Mountains, rivers, large bodies of water, deserts etc. offer effective barrier for movement of many insects especially wingless species. Ocean is the biggest barrier for thousands of species of insects. Inspite of availability of suitable areas, insects are unable to reach because of this factor alone.

Change of topography by man has also affected insects. With the increase of population more land has to be brought under plough. Thus, jungles are blown up. All these actions of man are bound to displace insects and they are forced to shift elsewhere and take to new host plants or get perished.

(c) Naturally Occurring Natural Enemies

(i) Parasites and Predators

There is hardly any insect which does not have natural enemy. Parasitisation of insects by insects themselves is very common.

Trichogramma spp. are egg parasites and are capable of parasitising more than 50 lepidopterous pests. Maize stem borer *Chilo partellus* larvae are parasitised by many parasites viz. *Apanteles colemani, Bracon chinensis, Stenobracon desae.* Because of the smaller size, parasites generally get annihilated by indiscriminate use of insecticides and their role as natural control agents gets minimized.

The beneficial job done by adults and grubs of predaceous coccinellids is well known to all. They take heavy toll of aphids. In addition, arachnids, fishes, amphibians, reptiles, birds and some mammals also play very important role in keeping the insect population in check. Early rising birds like Myna and Seven Sisters feed on cutworms which have not been able to hide themselves in soil by that time. These birds also probe into the soil to get the cutworms hidden therein. Spiders, amphibians, and some reptiles survive mainly on insects as food.

(ii) Pathogens

Microorganisms are known to cause diseases amongst insects. Some of the important insects which are affected by them are: *Ostrinia nubilalis* by *Glugea pyrasutae* (Protozoa); *Popilia japonica* by *Rickettsiella populliae* (Rickettsiae), *Chilo partellus* by *Beauveria densa* (Fungus), *Pieris brassicae* by *Bacillus thuringiensis* (Bacteria), *Spodoptera litura* and *Heliothis armigera by* Nuclear Polyhedrosis Viruses and *Tryporyza insertulus* by DD/136 (Nematode).

B. APPLIED CONTROL

When harmful insects are not held in check by natural factors, man has to take measures to bring down the population of insects.

(a) Cultural Control

Amongst the applied control methods, cultural methods are the oldest, least expensive, effective, need no complicated implements and no hazard to any animal life is involved.

(i) Stimulation of Crop Growth to Reduce Pest Incidence

Generally crops with exceptionally healthy growth have more insect attack because of having more succulent leaves. This however , does not hold good always. To give an example the sugarcane whitefly, *Aleurolobus barodensis* is more severe in the sugarcane crops which are deficient in nitrogen, the infestation can be checked by stimulating crop growth by applying nitrogenous fertilizer.

A mixture of "neem" and "mahua" *(Madhuca indica)* cakes if applied to sugarcane crop stimulates growth of the crop and keeps termites away from the crop.

(ii) Pruning and Thinning

Pruning has been proved to be very effective against the 'phalsa' bug, *Gargara mixta*. Thinning helps in the vigorous crop growth and saves the crop from damage by this pest.

(iii) Use of Resistant Varieties

This is one of the most important cultural measures in which if a resistant variety can be evolved against a particular pest, the infestation of atleast that particular pest is avoided. However, the resistant variety approach has some limitations. This needs regular checking up as sometimes the resistance is reversed. Sugarcane varieties Co 313, Co 421 and Co 527 have been found resistant to certain sugarcane borers. Paddy variety IR 20 is resistant to the green leaf hopper.

(iv) Crop Rotation

Crop rotation has been found to be very effective towards pests on large scale. A certain pest complex would attack only a particular group of plants and not all plants, so if the cropping scheme is such that after a particular crop has been harvested, another such crop may be sown whose pest complex may be different from that of the crop harvested earlier, the insects/insect stages which may be taking shelter nearby or may be hiding in the field itself may die because of lack of its host /alternative host plant and thus, may not attack the crop sown after the earlier crop has been harvested e.g., after pulses no leguminous crop may be sown. It may be followed by graminaceous crop. Crop rotation has been found to be effective against gram cutworm *(Agrotis* sp.) and cotton bollworms. Some of the cropping sequences suggested are given in Table 8.1.

(v) Ploughing and Tilling of Soil

By tilling the soil at regular intervals different hibernating stages of insects get exposed to sun and natural enemies like insectivorous birds. Tilling of soil is very effective in exposing and consequent destruction of the pupae of *Dacus cucurbitae.*

In general, hot weather cultivation helps in destroying the numerous stages of different insects. Fields all around paddy usually have vegetation which gives shelter to bugs and when the paddy reaches the milky stage about mid-August, the gundhy bugs migrate to paddy fields suck the milky sap from the grains and reduce to mere husk. Thus, if the weeds in the neighbourhood of paddy fields are cleaned sufficiently before the paddy grains reach milky stage most of these bugs would die because of loss of their shelter (the weeds around paddy fields).

The 'phadka' grasshopper, *Hieroglyphus nigrorepletus* lays eggs on the sides of bunds. Eggs are laid in a pod in vertical position. It has

TABLE 8.1: Some of the cropping sequences/rotations followed in India

1.	Paddy — Paddy	25.	Kharif onion—Rabi onion — Maize
2.	Paddy — Wheat	26.	Cucurbits — Tomato/Brinjal
3.	Maize — Wheat	*27.	Black gram — Garlic — Cucurbits
4.	Pearlmillet — Wheat	*28.	Cucurbits — Lentil — Okra (early)
5.	Sorghum — Wheat	*29.	Jute — Paddy
6.	Pearlmillet — Sorghum	30.	Paddy — Wheat — Maize
7.	Paddy — Fingermillet	31.	Paddy — Paddy — Groundnut
*8.	Groundnut — Wheat	*32.	Paddy — Groundnut
*9.	Pigeonpea — Mustard	*33.	Jute — Paddy — Potato
10.	Paddy — Paddy — Paddy	*34.	Paddy — Wheat — Jute
11.	Maize — Fingermillet — Maize	*35.	Paddy — Potato — Wheat
12.	Sorghum — Fingermillet — Maize	36.	Paddy — Sorghum — Paddy
*13.	Maize — Potato — Wheat	37.	Paddy — Fingermillet — Pearlmillet
14.	Jute — Paddy — Wheat	38.	Sorghum — Wheat — Green gram
15.	Sorghum — Wheat — Sorghum	*39.	Sorghum — Potato — Green gram
*16.	Sorghum — Cucurbits	*40.	Paddy — Pigeonpea — Maize
*17.	Groundnut — Mustard — Green gram	*41.	Pigeonpea — Wheat
*18.	Maize — Groundnut	*42.	Paddy — Jute
*19.	Maize — Potato	*43.	Paddy — Potato — Green gram
20.	Pearlmillet — Fingermillet	*44.	Cotton — Potato
*21.	Paddy — Cotton	45.	Maize — Wheat — Moong
*22.	Okra — Brinjal — Cucurbits	46.	Paddy — Paddy — Green gram
*23.	Paddy — Rabi maize — Green gram	47.	Jute — Paddy — Paddy
*24.	Kharif onion — Wheat/Mustard — Green gram	48.	Paddy — Wheat — Paddy

* Crop sequences relatively less prone to pest attack.

been proved experimentally that if the position of egg pods is altered by ploughing, 90 to 95 per cent eggs do not hatch out even if they are not picked up by birds from the ploughed soil. Locust swarm lays eggs, if the field is fallow, it can be ploughed to expose the locust eggs to birds. This alone is generally enough and no additional control measures may be needed.

(vi) Shifting the Sowing Time

By manipulating the time of sowing, the pest incidence can be reduced/checked. Hessian fly of wheat has a fall generation in September-October and if wheat is sown after the expiry of this fall generation, wheat crop is saved from the attack. Thus, a 'fly free week' is announced in different parts of U.S.A. for sowing the wheat crop.

Aulacophora foveicollis attacks cucurbits in January-February in the very early stages and the growing creepers get killed. It has been however observed that if the cucurbit seeds are sown a little earlier, the creepers are saved as by the time this beetle attacks, the creepers become hard enough to get killed by the attack.

During 1972-73 several IARI wheat fields were found attacked by the shoot fly *Atherigona naqvii* Steyskal. The crop sown either early before mid-October or late after December was found to be attacked by this pest but the crop sown during normal sowing season i.e., late October or early December was found to be free from shoot fly damage (Singh *et al.*; 1973)

The early season incidence of shoot fly resulted in 13.3 per cent and 1.2 per cent damage to wheat tillers in the crop sown on 10th and 20th October, respectively. The late season incidence was found to range from 56.5 per cent to 85.7 per cent in different January sowing dates (1st, 7th, 11th, 15th, and 20th). The shootfly incidence seemed to increase with delay in sowing date. Such a high incidence is a serious threat to late sown wheat crop (Singh *et al.*, 1973).

(vii) Clean Cultivation

This is an important item of cultural control measure. There are instances where clean cultivation has been able to control the pest.

(b) Mechanical Methods

Collection and destruction of egg masses mechanically prevents hatching and subsequent attack of the pest. In case of *Pyrilla* and *Leptocorisa* clipping of leaves with egg masses and their destruction is a common prevalent and effective practice. Hand picking of the pest wherever the pest is not an active flier is also an effective mechanical method of pest control.

(c) Physical Methods

Some of the common examples of physical methods of control are:

(i) Cotton seeds are sun-dried to kill the hibernating pink bollworm larvae.

(ii) Wheat and other grains are sun-dried or oven-dried to reduce the moisture to a certain level before storing to avoid insect infestation e.g., wheat is dried to reduce the moisture to about 8 per cent level and maize to about 10-12 per cent level.

(d) Chemical Control

(i) Synthetic Insecticides

Chemical insecticides are still the main defence against vast majority of insect pests. Such chemical insecticides may be inorganic compounds, synthetic organic compounds or organic compounds of plant origin and they may be used in the form of dust, spray, granule, aerosol, fumigant or bait.

Inorganic insecticides are mainly arsenicals (lead arsenate, calcium arsenate) and fluorine compounds (sodium fluoride, barium fluoride etc.). These were used in the past as stomach poisons. With the advent of synthetic organic insecticides, these have now disappeared from the scene, mainly because of their high mammalian toxicity.

Discovery of the insecticidal properties of DDT during the Second World War, conclusively proved that synthetic organic compounds could replace several inorganic and plant origin insecticides. Most of the insecticides used today belong to this group of insecticides.

Shortly after the discovery of the insecticidal properties of DDT in 1939 many other chlorinated hydrocarbons e.g., endrin, aldrin, dieldrin, heptachlor, thiodon, BHC (benzene hexachloride) etc. were developed. All these act mainly as contact insecticides. These are slowly getting replaced by more powerful and less persistent organophosphorus insecticides. However, endosulfan is being widely used even today because of its being comparatively less persistent and practically non-toxic to bees and certain biological control agents of insect pests.

Organophosphorus (OP) insecticides also were developed during the Second World War and the first of the series was hexaethyl tetraphosphate (HETP). Some of the widely used organophosphates are malathion, methyl parathion, dichlorvos, monocrotophos, trichlorfon sumithion, diazinon, demeton, phosphamidon, mevinphos, dursban disulfoton, phorate, dimethoate etc. Most of the OP. compounds are powerful contact insecticides and also act as stomach poison and some have fumigant action too. Many are readily translocated in plant system.

In 1947, another class of insecticide, carbamate was developed. The first insecticide of the group was dimetan patented in 1952. Although carbamate insecticides differ chemically from OP compounds, the killing action is similar to the latter; fairly wide range of carbamate insecticides are now being produced. They are both contact and systemic in action. The carbamates which are currently in use are: carbaryl, aldicarb and carbofuran.

Synthetic pyrethroids are the latest class of insecticides developed on the basis of structure of natural pyrethrin I present in flowers of *Chrysanthemum cinerarifolium.* The first synthetic pyrethroid, allethrin was developed in 1949, under the trade name pynamin. This has, however, limited use especially in mosquito coils. Further synthesis and screening led to the development of several powerful pyrethroids which are unmatched in toxicity. Thus, LD_{50} of decamethrin to housefly is only 0.3 as against 4.0 of DDT. There are several other pyrethroids e.g., permethrin, fenvalerate, recemethrin etc. A new synthetic pyrethroid, indothion has been developed at the National Chemical Laboratory, Pune. Its activity has been found to be comparable to other known pyrethroids. Its activity is like the natural pyrethrin and has some antifeedant properties also. Inspite of the safety in the use of pyrethroids the high cost of their production, is the major hurdle in their use.

Continuous and intensive use of synthetic insecticides has led to the segregation of insect strains resistant to insecticides. About 500 insect species have already been reported to have developed resistance and in many cases cross resistance to several insecticides. Indian scenario has not been very bad as only about 10 insect species affecting agricultural crops are reported to have developed resistance against synthetic insecticides (Table 8.2)

In spite of several serious drawbacks, synthetic insecticides are likely to remain on the scene for many more years to come.

There are a few other chemicals also e.g., chemosterilants, attractants, deterrents, insect growth regulators (IGR) which also are used in pest management. However, their use is limited.

(ii) Insecticides of Plant Origin

While at present insect pests by and large are being managed by synthetic

TABLE 8.2: Pests against which development of resistance to insecticides has been reported in India

Sl. No.	Pest	Insecticide	Level of resistance
1.	'Singhara' beetle	DDT, BHC	Control failure
2.	Tobacco Capterpillar	BHC	7 fold
		Malathion	5.7 fold
		Pyrethrum	14.7 fold
		Lindane	16.3 fold
		Endosulflan	85.9 fold
3.	Diamond back moth	Fenitrothion	Control failure
		Malathion	
		DDT	
4.	American bollworm	DDT	70-303 fold
		Endosulfan	12.2-12.5 fold
		Monocrotophos	3.6 fold
		Cypermethrin	50-300 fold
		Fenvalerate	120-287 fold
5.	Mustard aphid	Endosulfan	13-24 fold
		Malathion	4-6 fold
		Dimethoate	4-6 fold
6.	Flour beetle	DDT	2.3-10 fold
		Lindane	Resistant*
		Malathion	16-37.8 fold
7.	Rice weevil	Malathion	5.8 fold
		Lindane	Resistant*
		Phosphine	-do-
8.	Lesser grain borer	Malathion	-do-
		Lindane	-do-
		Phosphine	-do-
9.	Saw-toothed grain beetle	Malathion	-do-
10.	Khapra beetle	Phosphine	40 fold

* Precise level not determined.

insecticides, it is strongly felt that in future we may not be able to use these to the same extent, partly due to their ever increasing cost and partly due to the hazards involved in their use.

Plant origin insecticides are the oldest insecticides used by man for protection of crops and animals against insects and other arthropods. Though a number of plants were identified in the past, only a few such as nicotine from *Nicotiana tabaccum,* rotenone from *Derris elliptica* and pyrethrum from *Chrysanthemum cinerariifolium* could gain commercial importance. Of these, only pyrethrum survived and the rest were slowly pushed out of market by synthetic insecticides. Pyrethrum is still being used and is considered as an ideal insecticide because it is relatively much safer than synthetic insecticides, non-persistent, and at the same time exhibits high toxicity against insect pests. However, its high cost prohibits its large scale use and is, therefore, at present restricted for use against household, and fruit/vegetable pests.

Inspite of the presence of large number of synthetic insecticides, nicotine continued to attract scientists (Srivastava, 1953; Attri, 1978) because of certain plus points associated with its use: (1) it is highly volatile and thus, leaves no residue on treated surface, (2) it has triple action (stomach,contact and fumigant) against insect pests and (3) large quantity of tobacco waste (about 70,000 tonnes per year) is available and this has been identified by National Committee of Science and Technology (Anonymous, 1976a) as potential source of insecticidal nicotine sulphate.

Plant origin insecticides have a number of appealing qualities: (1) unlike synthetic insecticides, plant origin insecticides are renewable (2) these are much less hazardous and in certain cases safe e.g., neem, (3) though may appear costly would ultimately be cheaper than synthetic insecticides, (4) farmers can grow their own insecticide for use. Looking into the advantages of plant origin insecticides the latest emphasis has been to 'treck back' to plant origin insecticides for management of insect pests.

The only drawback considered to be associated with plant origin insecticides is their low persistency. This is however, a boon especially when we treat vegetable and fruit crops.

Plants are store house of chemicals and many of these are elaborated for defence against insect pests because the overall pressure of insects on the plants is much more than any other group of animal. Of the various mechanism developed by plants to defend themselves against insect pests, the role of chemicals elaborated by plants is considered most important. Such chemicals have now been identified as hormones, antihormones, antifeedant, repellent, oviposition deterrent, chemosterilant toxicants, etc.

Grainge and Ahmed (1988) have reported approximately 2400 plant species having pesticidal chemicals. Of these 1053 have been reported to possess chemicals having insecticidal activity. Many of these are reported to have diverse biological effects. For example, *Parthenium hysterophorus,* a weed occurring in India has antifeedant (Sharma and Hebbalkar, 1978) repellent and juvenile hormone (Nagasampagi *et al.*, 1982) and toxic property. Similarly root oil of *Acorus clamus* another indigenous plant is highly attractive to female Mediterranean fruit fly *Ceratitis capitata,* female melon fly, *Dacus cucurbitae* and female fruitfly *Dacus dorsalis* (Jacobson, 1976). Oil vapour of the plant is reported to sterilize male *Musca domestica* (Mathur and Saxena, 1975) and female *Thermobia domestica* (Saxena and Rohdendorf, 1974). These effects are in addition to its established toxicity against many insects. There are several other commonly occurring plants e.g., *Ageratum conyzoides* (Sharma *et al.*, 1977), *Lantana camara* (Singh, 1983). *Melia azedarach* (Teotia and Tewari, 1971), *Pongamia glabra* (Parmar and Gulati 1969). *Tagetes* (Singh and Kataria, 1985 Singh *et al.* 1987) *Hymenocallis littoralis* (Singh and Pant, 1980 a,b), *Annona squamosa* (Mariappan and Saxena, 1984) *Crinum bulbispermum* (Singh, 1978) etc. may have potential of practical utilization.

Neem *(Azadirachta indica)*

Of the 1000 odd plants so far reported to be insecticidal only 2-3 dozen plants at present appear to be of practical utilization and only one plant, *Azadirachta indica* (Syn. *Melia indica, M. azadirachta)* commonly known as neem could be commercialised. For centuries neem tree has been held in esteem by Indian folk because of its medicinal and insecticidal values. Mixing of neem leaves with grains during storage and keeping dried leaves between folds of clothes to protect against insect pests during storage is rather well known to Indian masses. A break through in the utilization of neem as source of insecticide was made by Pradhan *et al.* in 1962 at the Indian Agricultural Research Institute, New Delhi who for the first time reported that the active material is concentrated in the seed kernel. Aqueous suspension prepared from seed kernel was found to inhibit completely the feeding of desert locust, *Schistocerca gregaria* (Fig. 73-A). The potential of the aqueous suspension was later demonstrated during the same year under actual field condition when locust invaded Delhi.

FIG. 73: (A) Treated: Effect of neem seed kernel suspension spray on feeding of maize leaves by the desert locust. (Courtesy: K.P. Srivastava)

FIG. 73: (B) Untreated

Researches conducted during the last 10-15 years have demonstrated diverse behavioural and physiological effect of neem on insects: repellent (Jacobson, 1981), antifeedant (Pradhan *et al.*, 1962), oviposition deterrent (Singh and Srivastava, 1983), growth inhibition and production of malformed adults (Leuschner, 1972), sterilant (Dorn *et al.*, 1987) and also direct toxicity (Singh *et al.*, 1988).

About 40 compounds have been isolated from seed kernel which show biological activity against insects. Of these azadirachtin is considered the most active.

Neem extracts and pure compounds have been evaluated against more than 300 species of insects and almost all have been found susceptible. These include important key insect pests of agriculture e.g., *Schistocerca gregaria, Heliothis armigera* (Fig. 74), *Spodoptera litura, Dacus cucurbitae, Earias* sp., *Pectinophora gossipiella, Lipaphis erysimi, Nilaparvata lugens, Microtermes, Trogoderma granarium, Callosobruchus maculatus* etc.

A commercial neem formulation developed in USA is at present being sold under the trade name Morgosan-O (R). In India, at least 20 neem formulations, viz., 'Well grow', 'Neemark' 'Neem guard', Neemgold, Margocide OK, Margocide CK, Achook and RD-9 Replin are currently being marketed for use against insect pests. Several neem formulations are in pipeline and are likely to hit the market soon as effective substitute to synthetic insecticides.

There are about two crores neem trees in India with yield potential of about 5,00,000 tonnes of seed every year (Ketkar, 1976). Thus there is enough material available for management of insect pests. Availability, renewability and effectiveness make neem seed an ideal source of insecticide.

(iii) Safe Use of Pesticides

While on the one hand chemical pesticides are indispensible, on the other hand there is serious concern of contamination of environment and danger to warm-blooded animals. Various ways by which contamination of the environment due to the use of pesticides/insecticides could be minimised are given below:

(1) Need-based Use

Chemical pesticides should be applied only when these are really needed.

(2) Timing of Application

Application of the chemicals should be so timed that it may strike at the weakest and the most vulnerable stage in the life-cycle of the pest, e.g. borers could be more successfully killed by synchronising the application of insecticides with the peak period of hatching. Use of life tables prepared for the host crop and the related pest may be extremely useful in determining the right time for application of pesticides. Application of pesticides may also be so timed that the natural enemies are least affected.

(3) Use of Improved Techniques of Application

Quantitative reduction in the use of pesticides could be achieved by careful selection of the method of application. Treatment of seeds of sorghum (Jotwani, 1971) and barley (Singh and Ramachandran, 1981) with carbofuran prior to sowing has been found to be very effective for controlling the sorghum shootfly and barley aphids respectively. Similarly roots of paddy seedling dipped in chlorpyriphos before transplanting has given effective control of the paddy gall midge, stem borer and leaf hopper upto 30 days (Kalode, 1976). These methods of application of chemicals reduce the hazard to the minimum. Certain pests of trees have been controlled successfully by injecting pesticides instead of by the usual method of dusting and spraying. This technique affords protection of natural enemies also and has been used successfully for control of *Nephantis sirinopa,* the black headed caterpillar of coconut (Nadarajan and Channabasavanna, 1978) and mango hopper (Sandhu *et al.,* 1979).

Spot application of insecticides instead of treating whole crop in certain situations is far less hazardous. This is applicable in crops like maize. A few lindane granules in central whorl have given effective control of the maize borer.

(4) Use of Physiologically Selective Insecticides

Selection of insecticides may be made on the basis of susceptibility of the pest species and their developmental stages. Susceptibility should first be determined by laboratory trials before taking to field. Endosulfan has been found to be comparatively safer to the honey bee, *Apis indica* (Attri and Sharma, 1969) and can be sprayed on the crops without any detrimental effect on bees. Spraying during the afternoon hours is still better. Endosulfan has been found to be quite safe to adults of the parasite *Bracon brevicornis* (Sarup *et al.,* 1971).

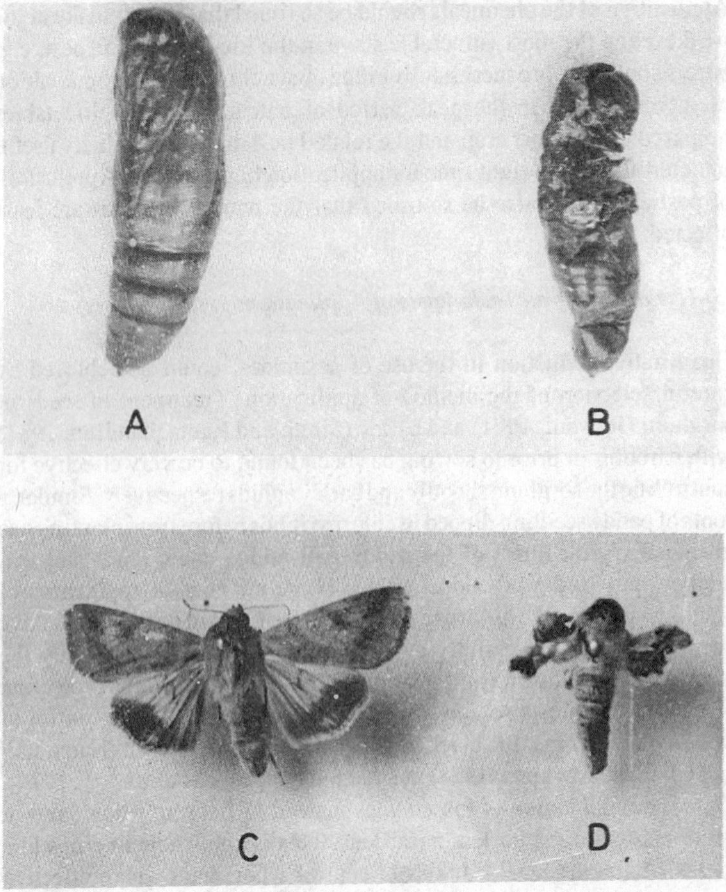

FIG. 74: Effect of Neem seed kernel suspension treatment on the growth of *Heliothis armigera* (Courtesy: K. P. Srivastava)

(A) Normal pupa formed from untreated larva

(B) Malformed pupa formed from the larva fed on neem seed kernel suspension-treated milky stage sorghum spikelets with grains

(C) Normal adult emerged from the pupa formed from untreated larva

(D) Malformed adult emerged from the pupa formed from the larva fed on neem seed kernel suspension-treated milky stage sorghum spikelets with grains.

(5) Use of Comparatively Safer Pesticides

Metcalf (1966) suggested the use of methoxychlor as a substitute for DDT. Malathion, ronnel, chlorothion are safer than most of the other organophosphorus compounds. Insecticides of plant origin are comparatively less persistent but much safer than synthetic pesticides. For the control of the codling moth, Ryania, an insecticide of plant origin, has been found to a suitable substitute for DDT. It gave a good conservation of natural enemies also (Pickett, 1959). Saxena *et. al.*, (1980) found 'Neem' (*Azadirachta indica*) to control the pests of paddy effectively whereas Singh *et al.*, (1985) found it to be effective against the pests of red gram (*pigeon pea*). Singh and Singh (1985) also found deoiled neem kernel powder to be very effective against *Trogoderma granarium.*

Pyrethrum is already known for its low mammalian toxicity and it is being used for the control of several pests including the stored grain pests.

(e) Biological Control

Biological control is the utilization of biotic factors (Parasites, Predators and Pathogens) for the control of insect pests. This method of insect pest management is probably the safest of all the methods known. More than 110 species of insect pests have already been managed to a greater or lesser degree (Paul De Bach, 1964).

Breeding of the egg parasites *Trichogramma* spp. (Fig. 75) and *Bracon* spp. is going on in different entomological laboratories in India. At the Indian Agricultural Research Institute (IARI), New Delhi, breeding of parasites was started more than four decades ago and is still going on. Some releases have also been made. At the Directorate of Plant Protection, Faridabad (Haryana), also breeding of parasites was started a few years back.

Biological control of insect pests may be done either by inundative releases of parasites/predators, after mass breeding in the laboratory or by the establishment of exotic parasites/predators in the field. In Maharashtra, success has been achieved by inundative releases of the egg parasite *Trichogramma chilonis* and 60-80 per cent larvae of the sorghum shoot fly, stem borer and *Heliothis* could be controlled. For effective control of the shoot fly 1-1.5 million adults of the parasite per hectare were required, whereas in case of the stem borer and *Heliothis*, 1.5-2.5 million adults of the parasite per hectare were needed (ICAR

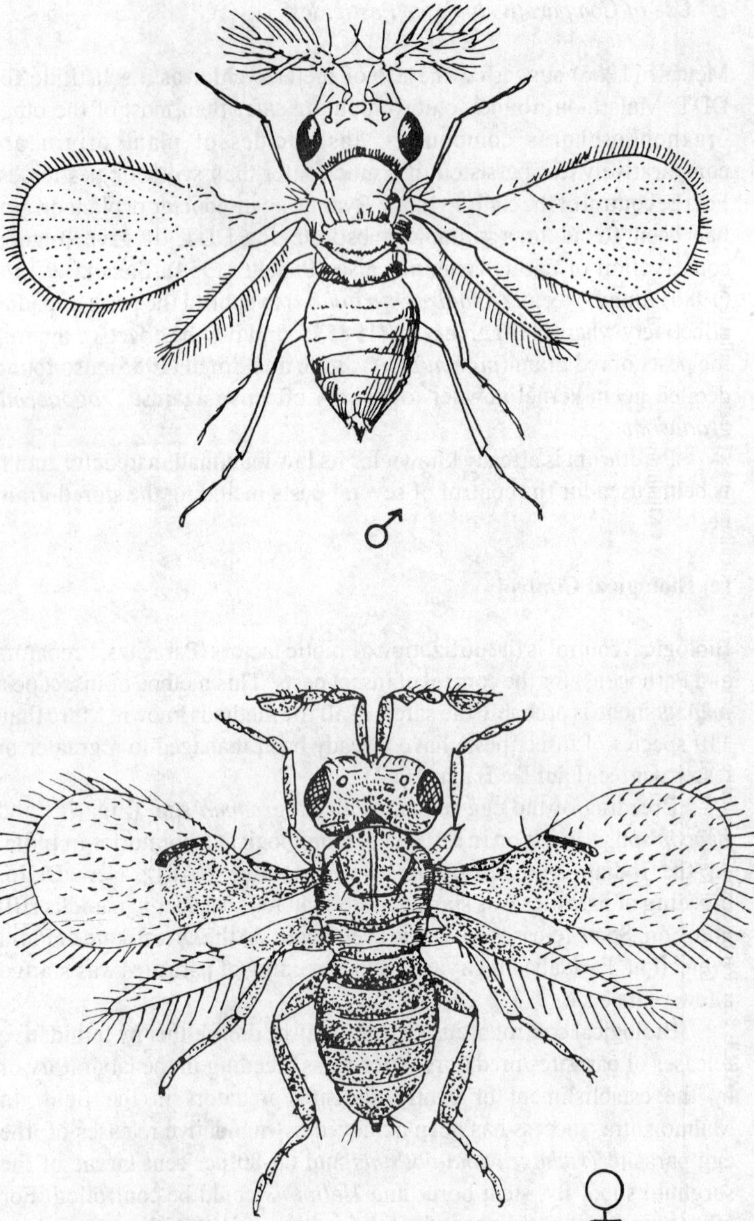

FIG. 75: Adults of *Trichogramma* sp.

TABLE 8.3: Parasites and Predators of agricultural and horticultural insect pests, available at Biocontrol Research Laboratories, Bangalore

Crop/Pest		Control agent parasite/predator	** Minimum release/hectare	* Price
Sugarcane				
Early shoot borer	: *Chilo infuscatellus*			
Internode borer	: *C. indicus*	Egg parasite	1,00,000	Rs. 40 for
Stalk-borer	: *C. auricilius*	*Trichogramma chilonis*		20,000 (i.e.
Plassey borer	: *C.tumidicostalis*			one 'Tricho card')
Gurdaspur-borer	: *Acigona steniellus*			
Top shoot-borer	: *Scirpophaga excerptalis*	*Trichogramma japonicum*	1,00,000	Rs. 40 for 20,000 (i.e., one 'Tricho Card')
Sugarcane scale	: *Melanaspis glomerata*	Predacious ladybird beetle		
		Chilocorus nigritus	1,500	Rs. 60 for 100
		Pharoscymnus horni	1,500	Rs. 60 for 100
Coconut				
Coconut black-headed caterpillar		Larval parasites		
Opisina arenosella		*Goniozus (=Parasierola) nephantidis*	3,000	Rs. 30 for 100
		Bracon brevicornis	3,750	Rs. 25 for 100
Coffee/Citrus/Grape/Mango/etc.				
For all mealybugs		Predacious ladybird beetle		
Planococcus,Maconellicoccus etc.		*Cryptolaemus montrouzieri*	1,500	Rs. 60 for 100
		Scymnus coccivora	1,500	Rs. 60 for 100
Citrus/Coffee mealybug		Parasite		
Planococcus citri		*Leptomastix dactylopii*	2,500	Rs. 40 for 100

TABLE 8.3 *(Contd.)*

Crop/Pest	Control agent parasite/predator	** Minimum release/hectare	* Price
Cotton			
Cotton bollworms	Egg parasite		
Earias vittella and *E. insulana*	*Trichogramma chilonis*	2,50,000	Rs.40 for 20,000 (i.e. one 'Tricho card')
Pectinophora gossypiella	Predacious green lacewing		
Helicoverpa (=*Heliothis*) *armigera*	*Chrysoperla carnea*	2,500	Rs. 20 for 100 eggs/larvae

* (i) Excluding packing, forwarding charges etc. Prices subject to change without notice. (ii) December, 1992 rates.

** Repeated releases may be required depending upon the prevailing pest population.

Annual Report 1988-89, 1989-90). Inundative releases of *T. brasilensis* in Ludhiana district (Punjab) are reported to have reduced the cotton bollworm incidence from 70 to 30 per cent.

A list of parasites and predators which are being mass bred in India for inundative releases, names of the pests against which these can be used and the number required to be released per hectare are given in Table 8.3.

A few examples of successful introductions of exotic parasites and predators in India are given below :

Cottony cushion scale *Icerya purchasi,* has been very successfully controlled by the lady bird beetle *Radolia cardinalis* in Kerala (India). This predator was introduced from California.

San Jose Scale, *Quadraspidiotus perniciosus,* a serious pest of apple has been controlled by *Prospaltella permciosi* and *Aphytis diaspidis* introduced from California and are reported to have achieved high level of parasitisation of San Jose Scale in Uttar Pradesh and Kashmir (India).

Wooly aphis, *Erisoma lanigerum* another serious pest of apple has been successfully controlled in Assam (India) by the parasite *Aphelinus mali.* This parasite was introduced in about 30 countries including India from Central America.

Telenomus remus, an egg parasite of *Spodoptera spp.*, was introduced in India from New Guinea and has established in Andhra Pradesh and Karnataka.

Two braconid parasites *Bracon kirkpatricki* and *Chęlonus blackburnii* introduced in India for the control of cotton boll worm, were recovered in Hissar (Haryana) and *C. blackburni* is being recovered in other states as well. Another egg parasite *Trichogramma brasiliensis* which parasitises the eggs has been recovered in Hissar. However this parasite together with *C. blackburni* are credited with 8 to 15 per cent reduction in the damage to cotton bolls and 10 per cent reduction in damage to chick pea by *Heliothis armigera.*

Potato tuber moth, *Phthorimaea operculella* is a pest of potato tubers throughout the tropical and subtropical regions of the world. Two parasites *Copidosoma koehleri* and *Apantles subandinus* introduced from South America in several countries are reported to be most successful species for the biological control of potato tuber moth.

Among the pathogens, bacteria and viruses have shown promise against several insect pests. Commercial preparations of *Bacillus thuringiensis* are available throughout the world. In India, it has offered some promise against the castor semilooper, *Achaea janata.* Spectacular success has been achieved by the use of nuclear polyhedrosis virus (NPV) in India and in many countries against a large number of lepidopterous pests. In India (Tamil Nadu and Andhra Pradesh) NPV extract concentrate is distributed for the control of *Amsacta albistriga, Heliothis armigera* and *Spodoptera litura.* Field trials conducted using NPV for the control of *Spodoptera litura* (Ramakrishnan *et al.,* 1981; Krishnaiah *et al.,* 1985) and *Heliothis armigera,* the gram caterpillar (Narayanan, 1979) have been found to be effective.

Biological control methods have the limitation generally, of not being used along with the chemical control agents and thus generally do not give immediate relief to the cultivator.

(f) Legislative Approach

Legislative methods are meant mainly to prevent the entry of exotic pests into the country and to prevent the spread of certain pests from one part of a state/country to another. In addition legislations are enacted from time to time for the control of certain pest epidemics.

These objectives are achieved by the enforcement of rules and regulations supported by legal sanctions for thorough examination of imported material at the port of entry, strict enforcement of quarantine regulations and disinfestation of material or its destruction, if necessary. To check the spread of particular pests from one part of the country to another domestic quarantine has to be enforced through the Destructive Insect Pests Act of 1914. The Govt. of India takes steps to prevent the entry and spread of exotic pests. The states are empowered to enact legislations for the control of pests within their jurisdiction, e.g., in Gujarat there is a law prohibiting the sowing of cotton seeds without prior drying under the sun. By doing this, the pink boll worm larvae hibernating in cotton seeds get killed before the seeds are sown.

Before the establishment of the Directorate of Plant Protection and Quarantine, staff of the Collector of Customs used to check the materials only at the sea ports. There was no arrangement to check the entry of infested material at the land frontiers. The entry of Plant material was prohibited although sometimes certain exceptions had to be made with the prior approval of the Government of India (Pruthi, 1969).

The Plant Protection and Quarantine Directorate, about four years after its establishment in 1949, established a well equipped Quarantine and Fumigation Station at Bombay. Later, quarantine stations were established at other important sea ports and two near Amritsar on the Indo-Pakistan border. Now there are arrangements for inspection and treatment of imported materials at the airports at Amritsar, Bombay, Madras, Calcutta and New Delhi.

As regards the import of plants and plant materials in India, it is permissible only at the ports /airports of Bombay, Madras, Calcutta, New Delhi, Amritsar and Vishakhapatnam and inspection stations have been established at these ports/airports.

To prevent the entry of certain exotic pests in the country import of a number of items from certain countries is prohibited. However, most of the plant materials can be imported if these are accompanied by health certificates issued by the exporting country. Import of useful insects e.g., exotic parasites and predators, honey bees and silk worms is possible with the special permission of the competent authority (Pruthi, 1969).

Different states of the country have passed agricultural pests, diseases and Noxious Weeds Acts to tackle the pests which can move from one port to another. States are empowered to declare an insect as an injurious organism.

To discharge the international quarantine obligations, the Government of India have framed certain rules and regulations (Anonymous, 1976). Some of these,briefly, are as follows:

(a) that all living plants, fruits and vegetables be imported through one of the specified seaports and must undergo inspection and, if necessary, treatment,

(b) that all imports of living plants and plant materials by air be routed through one of the specified airports and must also be accompanied by a special import permit issued by the Plant Protection Adviser to the Government of India,

(c) that all imports of plants and plant materials must be accompanied by an official phytosanitary certificate from the country of origin,

(d) that the import of plants and parts of plants of the genera *Allium, Citrus, Pinus, Ulmus* and coconut, cocoa, coffee, cotton, groundnut, potato, rubber, sugarcane and unmanufactured tobacco, is prohibited, except under certain specific conditions,

(e) that import of sugarcane from Australia, Fiji, New Guinea and Philippines, cocoa and other *Theobroma* species from Africa, Sri Lanka, and West Indies, rubber plants and seed from America and West Indies, sunflower seeds from Argentina and Peru, coconut from British Guiana, Carribean area, Ghana, Guam, Jamaica, Philippines, Togoland, and West Indies; unginned cotton and mexican jumping beans, is absolutely prohibited,

(f) that import of seeds of berseem, coffee, cotton, flax, rubber, *Theobroma* species and tobacco is prohibited, except under special authorisation from Research Organization or the State Department of Agriculture to the effect that the importation is for scientific purposes only.

In exercise of the powers to enforce domestic quarantine, the Government, through a series of notifications published in the Gazette of India (Anonymous, 1976b), have placed restrictions on the movement of certain articles which can help in the dispersal of notified pests and diseases, except through certain specified routes, to ensure,

(a) against the dispersal of fluted scale, *Icerya purchasi* Maskell, from Karnataka, Kerala and Tamilnadu (infected area) to the rest of the country,

(b) against the spread of San Jose Scale, *Quadraspidiotus perniciosus* Comstock, from J&K, Himachal Pradesh, Punjab, Uttar Pradesh, West Bengal, Assam, Orissa and Tamilnadu to the rest of the country,

(c) against the spread of banana bunchy top disease from Assam, West Benagl, Orissa, Tamil Nadu and Kerala to the rest of the country,

(d) against the spread of mosaic disease of banana from Maharashtra and Gujarat to the rest of the country, and

(e) against the spread of wart disease of potato caused by *Synchytrium endobioticum,* from West Bengal to the rest of the country. For this purpose, the movement of potato tubers from the infected areas to the rest of the country is banned.

The quarantine Rules and Regulations are enforced strickly to prevent the entry of exotic pests. A few years back a consignment of Hops rooted cuttings used in breweries was smuggled into India but was found heavily infested with cyst nematodes hence could not be allowed in the country. Consequently, the entire consignment was destroyed. However, there are instances when quarantine has been bypassed and certain pests have entered the country. List of a few such pests is given in Table 8.4 (Pers. Commun: B.R. Verma).

TABLE 8.4 : A few accidental introductions of pests in India from foreign countries

Common Name	Scientific Name	Host plants	Introduced from	Year of Ist record
San Jose Scale	*Quadraspidiotus perniciosus* (Coccidae : Hemiptera)	Fruit trees	China	1889
Fluted scale	*Icerya purchasi* Maskel (Coccidae: Hemiptera)	Fruit trees	Australia	1909
Potato tuber moth	*Phthorimaea* operculella Zell. (Gelechiidae: Lepidoptera)	potato	Italy	1900
Golden nematode	*Heterodera rostochinensis* (Nematoda)	potato	U.K.	1961

Note : There is another legislation of the Government of India, the Prevention of Food Adulteration Act 1954 as amended upto August 1, 1962. According to this act, mixing of pesticides with foodgrains is prohibited.

(g) Integrated Pest Management

The development of Integrated Pest Management has been the most recent chapter in the history of pest management (Flint and Bosch, 1981).

Integrated Pest Management (IPM) is an ecologically based strategy which relies very much on natural mortality factors like natural enemies and weather and devise a control strategy which may least disrupt these factors (Flint and Bosch, 1981).

The U. N. Food and Agriculture Organisation's Panel of Experts on Integrated Pest Control (1967) defined IPM as :

> "A pest management system that in the context of the associated environment and the population dynamics of the pest species, utilises all suitable techniques and methods in as complete a manner as possible and maintains the pest population at levels below those causing economic injury".

The Council of Environmental Quality (CEQ) in its publication of 1972 Integrated Pest Management (IPM) defined IPM as an approach that employs a combination of techniques to control the wide variety of potential pests that may threaten crops. It involves maximum reliance on natural pest population controls along with a combination of techniques that may contribute to suppression-cultural methods, pest-specific diseases, resistant crop varieties, sterile insects, attractants, augmentation of parasites or predators, or chemical pesticides as needed.

IPM programme takes into consideration all available pest control actions including no action and evaluates the potential interaction among various control measures, cultural practices, weather, other pests and crops concerned. Under IPM, natural enemies,cultural practices, crops resistant to pests, microbial agents, genetic manipulation, messenger chemicals (e.g., sex attractants) and pesticides become naturally augmentative instead of individually exclusive or even antagonistic as has been so often under the pesticide- dominated control. Flint and Bosch (1981) consider the following as the six basic elements of IPM:

(i) the people concerned i.e., the system devisers and pest managers,
(ii) knowledge and information necessary to devise the system and make sound management decisions,
(iii) programme for monitoring the numbers and state the ecosystem elements e.g., resource, pests and natural enemies,
(iv) decision-making levels i.e., the pest densities at which control measures are put into action,
(v) the techniques used to manipulate pest populations, and
(vi) agents and materials which are the tools of manipulation.

The persons who implement the IPM programme must constantly study the situation and evaluate pest populations and according to earlier experience and knowledge initiate action and choose appropriate materials and agents as conditions demand. Those implementing the IPM programme should be conversant with the ecological and biological effects of their actions not only on the target pest but also on the natural enemies and non-pest organisms.

IPM is an approach to reduce pest damage to tolerable levels through several techniques e.g., natural predators and parasites, genetically resistant hosts, environmental modifications and if necessary suitable chemical pesticides. IPM strategies generally rely first upon biological densities against pests before chemically altering the environment.

Boza-Barduccii (1965), in a paper presented at the 12th meeting of cotton commodity production investigation at Washington stated that in Peru prior to 1939 only six species were reported as pests and were controlled by the use of arsenicals, nicotine insecticide, by hand picking of insect pests and by destruction of infested parts of plants. Natural control was carried by predators and parasites. However, by 1949-50 the cultivators switched over to synthetic insecticides like DDT, BHC and toxaphene etc. The use of these gave a good control of major pests like *Heliothis virescens* and *Aphis gossypii* for about five years resulting in the increase of crop yield. Later it was observed that these insecticides were no longer so effective and they switched over to aldrin, dieldrin and endrin; frequency also of the treatments was increased. Later they switched over to ethyl/methyl parathion. Now they realised that the use of pesticides was neither very effective nor economical. Six new pests also appeared, several developed resistance towards the insecticides used and about 50 per cent of the crop was lost.

Thus after about 7-8 years of switching over to synthetic insecticides, in 1957, they switched over to integrated control programme which included reduction in the ratoon cotton crop to 1/4 , adopting cultural practices, reduction of the use of pesticides to the minimum and introduction of natural enemies of pests from other countries and getting those established. A few years later, the situation improved and natural balance was restored. This single example should be enough to understand that integrated pest management is good in all respects.

There are several other important ingredients of IPM e.g., chemosterilants, sex pheromones, sex attractants and light traps. Of these, sex attractants and sex pheromones are of special significance because they are specific. Trimedlure, cue lure and methyl eugenol are attractants for three species of fruit flies and are currently used for the control of the Mediterranean fruit fly, *Ceratitis capitata* Wiedemann the melon fruit fly, *Dacus cucurbitae* Coquillet and the oriental fruit fly *D. dorsalis* Hendel.

Several sex pheromones are already in use in India in the IPM for the management of *Sodoptera litura, Heliothis armigera, Pectinophora gossypiella* and *Earias* sp. Thus, there are several ways and means to make IPM effective and successful.

In India in the villages, farmers have been actually following the integrated control programme unconsciously since long. Summer ploughing to expose the pests hiding thereinto scorching sun, digging the stubbles and burning along with the borers hiding in the stubbles, removing weeds from the cropped fields at regular intervals, collecting and destroying eggs and other stages of pests which could be collected easily, have been the age-old cultural practices and are integral parts of integrated pest management. With the advent of organic pesticides, the ease with which these could be used and the quick results the farmers obtained, they started considering the cultural practices as outdated and laborious. Due to the frequent use of pesticides and their indiscriminate use, certain pests started developing resistance towards the organic pesticides with the result that higher doses and increased frequency became necessary to get effective control of the pests. Thus, it became more expensive and more hazardous. In some cases, change of pesticide also became necessary as even the higher dose of earlier ones no longer remained effective. With all these developments, the so-called outdated practices were being revived and side by side chemicals also are being used judiciously so that the use of chemicals may not be much hazardous and delay the development of resistance of the pests towards pesticides.

An international seminar on Integrated Pest Control was organised in 1969 at New Delhi. All were unanimous that integrated approach to pest problems is definitely superior to any other method used alone. It is cheaper, more efficient,would avoid the upsetting of the balance in nature, delay the development of resistance of pests to pesticides because of comparatively lesser use of pesticides and will minimise the residue hazards of pesticides (Pradhan,1983)

Thus, integrated pest management avoids upsetting the balance in nature, delays the development of resistance of pests to pesticides, and minimizes the residue hazard by integrating the use of pesticides, biological control agents, ecological and chemical control measures.

Agricultural Entomologists should conduct researches to find out the chemicals which may cause mortality of the pests but spare their natural enemies and pollinators etc. Thus, what is required is to investigate the chemicals which may be compatible with the biological control agents of pests and vice versa. If the scientists could locate such biological control agents which may be taxonomically away from the respective pests i.e., may not belong to the same phylum/ class/order to which the concerned pest belongs, these could be used simultaneously along with the chemicals which would kill the pest and side by side allow the biological control agents to do their job.

BIBLIOGRAPHY

Ananthakrishnan, T. N. 1969. Indian Thysanoptera, CSIR Zoological Monograph No. 1, New Delhi 171 pp.

Ananthakrishnan,T. N. 1971. Thrips (Thysanoptera) in Agriculture, Horticulture and Forestry—Diagnosis, bionomics and control. *J. Sci. Ind. Res.,* 30 (3) : 113-146.

Anonymous 1976a. Farm News. *Pesticides,* 10 (5) : 51.

Anonymous, 1976b. The Destructive Insects and pests Act, 1914 and Rules and Regulations framed thereunder (as corrected upto 1976). *Cyclostyled and circulated by the Dte.* Pl. *Prot. Quar. and Stor. Faridabad, Min. of Agr., Govt. of India:* 75 pp.

Anonymous, 1986. *Directorate of Plant Protection, Quarantine and Storage Organisation and Functions.* Issued by the Plant Protection Adviser to the Government of India, Faridabad, Haryana, 13 pp.

Anonymous, 1993. *Statistical Information on Plant Protection. Directorate of Plant Protection, Quarantine and Storage, Government of India,* Faridabad, Haryana, 33, pp.

Arora, R. K. and Sharma, S. 1990. Insects in the Biological Control of weeds. *Biodigest.* 4 (1 and 2) : 67-69.

Attri, B.S. 1978. Scope for improvement in formulation of insecticides of plant origion. *Proc. Meeting on Insecticides of Plant Origin,* New Delhi pp.133.

Attri, B.S. and Sharma, P.L. 1969. Toxicity of some new insecticides to Indian honey bee *(Apis indica* F.) *Pesticides,* 3 (6) : 27-29.

Atwal, A.S. and Sharma, O.P. 1970. Brood rearing of the Italian honeybee *Apis mellifera* in the Punjab plains at Ludhiana. *Indian Bee Jour.,* 32 (3-4): 62-67.

Ayyar, T.V.R. 1939. Entomology in India. *Indian J. Ent.* 1:9-16.

Ayyar, T.V.R. 1940. *Handbook of Economic Entomology for South India.* Govt. Press, Madras. 528 pp.

Banerjee, Barun Deb, 1983. Pests of tea. In: *Agricultural Entomology II* (Eds.) P. D. Srivastava, M.G. Jotwani, R.A. Agarwal, S.R. Wadhi, R.K. Bhanotar, R. K. Bhatnagar. All India Scientific Writers' Society. New Delhi, pp.261-276.

Bartlett, B.R. 1964. Integration of Chemical and Biological control. In: *Biological control of Insect Pests and Weeds* (Ed.), Paul De Bach, Chapman and Hall Ltd.,II, New Fetter Lane, London, pp. 489-511.

Batra, H.N.1983. Pests of temperate fruits. In: *Agricultural Entomology II.* (Eds.) P.D. Srivastava, M. G. Jotwani, R. A. Agarwal, S.R. Wadhi, R. K. Bhanotar, R. K. Bhatnagar. All India Scientific Writers' Society, New Delhi. pp. 212-234.

Beirne, B. P. 1967. *Pest Management.* Leonard Hill, London 123 pp.

Boericke, W. 1991. *Pocket Manual of Homoeopathic Materia Medica,* 1079 pp. B. Jain Publisher (Pvt.) Ltd., New Delhi.

Boza- Barduccii, T. 1965. Experiments about the utilization of the biological control methods and integrated control method of cotton pests in Cancle Valley, Peru (S.A.). Paper Presented at the *12th Meeting. Cotton Commodity Production Investigation,* Washington, D.C.

Butani, D. K. 1974. Pests damaging roses in India and their control. *Pesticides,* December: 40-42.

Butani, D. K. and Dadlani, N. K. 1983. Pests of Ornamental plants in India and their control. *Pesticides,* March, 18-20, 25.

Butani, D. K. and Srivastava, R. P. 1983. Insect pests of tropical and subtropical fruits. In: *Agricultural Entomology II* (Eds) P.D. Srivastava M. G. Jotwani, R. A. Agarwal, S.R. Wadhi, R.K. Bhanotar, R.K. Bhatnagar, All India Scientific Writers' Society. New Delhi. pp.189-211.

Carson, R. 1962. *Silent Spring.* Fawcett Publications, Inc., Greenwich, Conn., 304 pp.

Chapman, R.F. 1971. *The Insects: Structure and Function.* The English Language Book Society and the English Universities Press Ltd., London, 819 pp.

Conway, G.R. 1976. Man versus Pest. In: *Theoretical Ecology: Principles and Applications.*(Ed.) R. M. May. Backwell Scientific Publications, Oxford, pp. 257-281.

Crane, E. and Walker, P. 1983. The Impact of Pest Management of bees and Pollination. *Report pulished by the International Bees Research Association and Tropical Development and Research Institute.* London, 129 pp. and 2 annexes.

David, A. L. and Muthukrishnan, T. S. 1953. Summaries of *Opuntia* control by insects. The prickly pear cochineal: Observation on its natural hosts and enemies in South Africa. *Indian J. Ent.*, 15 (3) : 219-224.

Dorn, A., Rademacher, J.M. and Sehn, E. 1987. Effect of azadirachtin on reproductive organs and fertility in the larvae of milk weed bug,

Oncopeltus faciatus. Proc. 3rd. Int. Neem Conf., Nairobi, 1986, pp. 273-288.

D'Souza, G. I., Krishnamoorthy, Bhat, P. Venkataramaiah, G. H. and Chako, M. J. 1983. Pests of coffee. In: *Agricultural Entomology II* (Eds.) P. D. Srivastava, M. G. Jotwani, R. A. Agarwal, S. R. Wadhi, R. K. Bhanotar, R. K. Bhatnagar. All India Scientific Writers' Society, New Delhi, pp. 277-293.

Eisner, T., Hurst, J.J. and Meinwald, J. 1963a. Defence mechanisms of Arthropods. XI. The structure, function and phenolic secretions of the glands of a chordeumoid millipede and a carabid beetle. *Psyche*, 70 (2): 94-116.

Eisner, H.E., Eisner, T. and Hurst, J.J. 1963b. Hydrogen cyanide and benzaldehyde produced by millipedes. *Chem. and Ind.*, 124-125.

Fletcher, T.B. 1919. *Proc. 3rd Ento. Meeting,* Pusa (Bihar), India.

Flint, M. L. and van den Bosch, R.1981. *Introduction to Integrated Pest Management,* Plenum Press, New York and London, 240 pp.

Frost, S. W. 1942. *General Entomology.* McGraw-Hill Book Co., Inc. New York and London, 524 pp.

Gardner, J. C. M. 1944. A note on the imported Lantana bug (*Teleonemia scruphlosa* Stal.). *Indian Forester,* 70:139-140.

George Ordish, 1967. The economics of biological and chemical pest control In: *Biological Method in Crop Pest Control,* (Ed) George Ordish Constable and Co., Ltd., Orange Street, London, W.C.P. pp.187—217.

Gosard, G. A.1909. Relation of Insects to human welfare, *J. Econ. Ent.*, 2: 313—332.

Grainge, M. and Ahmed, S. 1988. *Handbook of Plants with Pest Control Properties.* John Wiley and Sons, New York, 470 pp.

Headley, J. C. 1972a. Defining the economic threshold. In: *Pest Control Strategies for the Future.* National Academy of Sciences, Washington pp. 100-108.

Headley, J. C., 1972b. Economics of agricultural pest control. *A. Rev. Ent.*, 17:273-286.

Hickman, C. P. 1966. *Integrated Principles of Zoology.* The C. V. Mosby Co. St. Louis and Toppan Co., Ltd., Tokyo, Japan, 965 pp.

Howlett, H. B. 1919. *Proc. First Mtg. Vet. Officers,* Lahore, India

Hughes, G. M. 1952. The coordination of insect movements. I. The walking movement of insects. *J. Exp. Biol.* 29:267-284.

Husain, M. Afzal, 1938. Entomology in India, Past, Present and Future, *Proc. 25th Indian Science Congress,* Calcutta, Presidential Address, 25: 201-246.

Imms, A. D. 1936. The ancestry of insects. *Trans. Soc. Brit. Ent.* 3:1-32.

Imms, A. D. 1965. *A General Text Book of Entomology.* Asia Publishing House, Bombay, Calcutta, New Delhi, Madras, London, New York. 886 pp.

Jacobson, M. 1976. Indian Calamus root oil: Attractiveness of the constituents to oriental fruit flies, melon flies and Mediterranean fruit flies. *Lloydia,* 39 (6) : 412-415.

Jacobson, M. 1981. Neem Research in U.S. Department of Agriculture: Chemical, Biological and Cultural aspects. *Proc. Ist Int. Neem Conf.* Rottach Egern. 1980, pp. 33-42.

Jayanth, K. P. 1987. Biological control of weeds in India. *Proc. Seminar cum Sixth Workshop on Biological Control of Crop Pests and Weeds.* Indian Institute of Horticultural Research, Hessaraghatta Lake Post, Bangalore. 560089.

Jayanth, K. P. and Nagarakatti, S. 1987. Host-specificity of *Neochetin bruchi* Huslache (Coleoptera: Curculionidae) introduced for biological control of water hyacinth in India. Paper presented at First *National Seminar on Biological Control,* Gujarat Agril. Univ. Anand, 29th June, 1987.

Johnson, W. H., Delanney, L. E., Williams, E. C. and Cole, T. A. 1969. *Principles of Zoology.* Holt., Rinehart and Winston, Inc. New York, U.S.A. 814 pp.

Jotwani, M. G. 1971. Seed treatment of sorghum for control of shootfly. In: *Control of Sorghum Shootfly* (Eds.) M.G.Jotwani, and W.R. Young, Oxford and IBH Pub.Co., New Delhi. pp. 262-273.

Jotwani, M.G. and Srivastava, K.P. 1984. A review of neem research in India in relation to insects. In: *Natural Pesticides from Neem tree* (*Azadirachta indica* A. Juss) *and other tropical plants.* (Eds.) H. Schmutterer, and K.R.S., Ascher, and F.R.G. Rauschholzhausen, pp. 43-56.

Jotwani, M. G., Sukhani, T.R. and Singh, Santokh, 1971. Seed treatment of sorghum for the control of shootfly. *Entomologists' Newslet.* 1 (10): 72.

Joy, P. J., Sathesan, N. V., Lyla, K. R. and Joseph, D. 1985. Successful biological control of floating weed, *Salvinia molesta* Mitchell using the weevil *Cyrtobagous salviniae* Calder and Sand in Kerala (India). Paper presented at *Asian-pacific Weed Sciences Society, Tenth Conference* Chiangmai. 24-30 pp.

Kaestner, A. 1968. *Invertebrate Zoology.II.* Interscience Publishers—A Division of John Wiley and Sons, New York, London, Sydney, 472 pp.

Kalode, B. B. 1976. Current status of research on rice pest control. *Pesticide Inf.* 2 (3): 40-44.

Kapil, R. P. and Jain, K. L. 1980. *Biology and Utilisation of Insect Pollinator for Crop Production.* Dept. of Zoology, Haryana Agricultural University, Hissar, India, 81 pp.

Ketkar, C. M.1976. *Final Technical Report: Utilisation of Neem and its byproduct,* (Entomology section) *Khadi and Village Industries Commission,* Bombay, India, pp. 117-164.

Krishnaiah, K., Ramakrishnan, N. and Reddy, P.C. 1985. Control of tobacco caterpillar, *Spodoptera litura* (F.) on black gram by nuclear polyhedrosis virus. *Indian J. Agric. Sci.* 55 (12): 775-776.

Leuschner,K. 1972. Effect of unknown plant substance on a shieldbug. *Naturwissenschaften,* (59): 217-218.

Lohumi,C. 1980. *Koori khane Wala Keera* (in Hindi) (Insect feeding on *Lantana).* Indian Counc. Agric. Res., New Delhi. 34. pp.

Mani, M.S. 1982. *General Entomology.* Oxford and IBH Publishing Co., Calcutta, Bombay, New Delhi, 912 pp.

Mariappan, V. and Saxena, R.C. 1984. Custard apple oil, neem oil and their mixtures: effect on survival of *Nephotettix virescens* and on rice tungro virus transmission. In: *Natural Pesticides from Neem Tree (Azadirachta indica A. Juss).* (Eds) H. Schmutterer and K.R.S. Ascher; Rauischholzhausen, FRG. pp. 413-429.

Mathews, G. A. 1984. *Pest Management.* Longman London and NewYork, 231 pp.

Mathur, A. C. and Saxena, B.P. 1975. Induction of sterility in male house flies by vapour of *Acrorus calamus L.* oil. *Naturwissenschaften,* (62): 576-577.

Menon, M. G. Ramdas, 1972. Dr. Bonemissza's Entomological Mission. *Entomologists' Newslet.* 2 (8) :53.

Metcalf, R. L. 1966. Metabolism and fate of pesticides in plants and animals. In: *Scientific Aspects of Pest Control.* Natn. Acad. Sci. U.S.A. Washington pp. 230-250.

Metcalf, C. L. and W.P. Flint, 1979. *Destructive and useful Insects—their Habits and Control.* Tata-McGraw-Hill Publishing Co. Ltd., New Delhi, 1087 pp.

Metcalfe, Z. P. 1940. How many insects are there in the world? *Ent. News* 51: 219-222.

Miller, D. F. and Haub, J. G. 1960. *General Zoology.* Holt., Richard and Winston, New York, Chicago, San Francisco, 550 pp.

Mohan, R. N. 1981. Parasites and Parasitic diseases. In: *Handbook of Animal Husbandry* (Ed.) P. L. Jaiswal. Indian Counc. Agric. Res., New Delhi, pp. 425-526.

Nadarajan, L. and Channabasavanna, G. P. 1978. Control of coconut black-headed caterpillar—*Nephantis serinopa* Meyrick (Lepidoptera: Cryptophasidae) by stem infection of systemic insecticide. *Curr. Res.,* 7: 25-26.

Nagasampagi, B.A., Gupta, A.S. and Sharma, R.N. 1982. Pest control agents of plant origin. *Proc. Indo-US Workshop on Biodegradable Pesticides*, Lucknow, 1979, pp. 285-291.

Narayanan, K. 1979. Studies on the nuclear polyhedrosis virus of gram caterpillar, *Heliothis armigera* (Hb.) (Lepidoptera: Noctuidae). Ph.D. Thesis, Tamil Nadu Agril. University, Coimbatore.

Parker, T. J. and Haswell, W. A. 1951. *A Text-book of Zoology I*. Macmillan and Co. Ltd., London, 770 pp.

Parmar, B. S. and Gulati, K. C. 1969. Synergists for pyrethrins (11)- Karanja. *Indian J. Ent.*, 31 (3) :239.

Paul De Bach, 1964. *Biological Control of Insect Pests and Weeds.* Chapman and Hall Ltd., London, 844 pp.

Pickett, A.D. 1959. Utilization of native parasites and predators. *J.Econ. Ent.* 52 (6): 1103-1105.

Pilllai, P.N.R. and Jayarathnam,K. 1983. Pests of rubber. In: *Agricultural Entomology II.* (Eds.) P.D. Srivastava, M.G. Jotwani,R.A. Agarwal, S.R. Wadhi, R.K. Bhanotar, R.K. Bhatnagar. All India Scientific Writers' Society, New Delhi, pp.310-318.

Pradhan, S. 1961. Probable role of biotic factors in the periodicity of locust cycles. *Indian J. Ent.*, 23 (1):1—6.

Pradhan, S. 1969. Integrated Pest Control: Philosophy and Feasibility. *Pesticides,* April, 1969: 15-24.

Pradhan, S. 1973. Plant Breeding through the window of pest control. *Entomologists' Newslet.*, 3 (2): 12—17.

Pradhan, S. 1983. Integrated Pest Control. In: *Agricultural Entomology II.* (Eds) P. D. Srivastava, M. G. Jotwani, R. A. Agrawal, S. R. Wadhi, R.K.,Bhanotar,R.K. Bhatnagar. All India Scientific Writers' Society, New Delh, pp. 420-434.

Pradhan, S. and Jotwani, M.G.1964. History of Entomology in India. In: *Entomology in India* (Ed.) N.C. Pant. Entomological Society of India, New Delhi, pp. 1-16.

Pradhan, S., Jotwani, M. G. and Rai, B. K. 1962. Neem seed deterrent to locusts. *Indian Fmg.*, 12 (8) :7-11.

Prasad, V. G., Krishnaiah, K. and Jagan Mohan, 1983. Insect pests of vegetable crops. In: *Agricultural Entomology II.* (Eds.) P. D. Srivastava, M. G. Jotwani, R. A. Agarwal, S. R. Wadhi, R. K. Bhanotar, R. K. Bhatnagar. All India Scientific Writers' Society, New Delhi. pp. 235-260.

Pruthi, H. S. 1969. *Text Book of Agricultural Entomology.* Indian Counc. Agric. Res., New Delhi, 977 pp.

Ramakrishnan, N., Chaudhary, S., Kumar, S., Rao, R.S.N. and Satyanarayana,

S. V. V. 1981. Field efficacy of nuclear polyhedrosis virus against tobacco caterpillar, *Spodoptera litura (F.)* on Tobacco. Tob. Res. 2 (2): 129-132.

Randhava, M. S. 1978. *Report of the Review Committee on Agricultural Universities.* Indian Counc. Agric. Res., New Delhi.

Randhawa, M. S. 1979. *A History of the Indian Council of Agricultural Research.* Indian Counc. Agric. Res., New Delhi, 510 pp.

Richards, O. W. and Davies, R. G. 1977. *Imms' General Text Book of Entomology.* Chapman and Hall, London and New York.

Roonwal, M. L. 1983. Termites of Agricultural Importance. In: *Agricultural Entomology II.* (Eds) P. D. Srivastava, M. G. Jotwani, R. A. Agrawal, S. R. Wadhi, R. K. Bhanotar, R. K. Bhatnagar. All India Sci. Writers' Society, New Delhi, pp. 23-35.

Sandhu, G. S., Sohi, A. S., Batra, R. C. and Bhalla, J. S. 1979. Stem infection for the control of mango pests with particular reference to mealy bug, *Drosicha mangiferae* Green and *Amritodus atkinsoni* L. Research Report on Mango Tech. Dec. 13, Mango Workers Meeting Panji, Goa, May 1979.

Sardar Singh, 1964. Bee keeping. In: *Entomology in India* (Ed.) N. C. Pant. Entomological Society of India, New Delhi, 497-511 pp.

Sarup, P., Singh, D. S. Srivastava, M.L. and Singh, R.N. 1971. Laboratory evaulation of pesticides against adult of an important parasite, *Bracon brevicornis* Wesmael (Hymenoptera: Braconidae) of some crop pests. *Indian J. Ent.,* 33 (3): 346-349.

Saxena, B. P. and Rohdendorf, E. B. 1974. Morphological changes in *Thermobia domestica* under the influence of *Arorus calamus* oil vapour. *Experientia,* 30:1298.

Saxena, H.P. 1974. Severe and widespread occurrence of *Maruca testulalis* Geyer in red gram, *Cajanus cajan. Entomologists' Newslet.* 4 (3) : 21.

Saxena, R. C. 1987. Thrips of Economic importance and their Management. In: *Recent Advances in Entomology.* (Eds.) Y. K. Mathur, N. D. Pandey, K. D. Upadhyay, J.P. Srivastava, Gopal Prakashan, Kanpur, pp. 141-168.

Saxena, R. C., Liquido, N. J. and Justo, H. D. 1980. Neem seed oil, a potential antifeedant for the control of the rice brown plant hopper, *Nilaparvata lugens,* In: *Natural pesticides from the neem tree (Azadirachta indica* A. Juss) (Eds.) H. Schmutterer, K. R. S. Ascher and H. Rembold, Rottach-Egern, pp. 171-188.

Sharma, R. N. and Hebbalkar, D. S. 1978. Pheromone and age-dependent correlates of sexual activity and oviposition in hemipteran bug. *Proc. VII. Annual Conf. Ethological Soc. India,* RB 12/25.

Sharma, R. N., Nagasampagi, B. A., Kasturi Rangachar, S., Hebbalkar, D. S. and Joshi, V. N. 1977. Investigations on the occurrence and role of anti-allatotropius in some plants and insects and their effects on insect development and Host-Plant relationship. Symposium on *Insect and Environment,* Delhi University, Abstracts, pp. 30-52.

Sharma, V. P. 1990. The relevance of Ethnomedicozoological drugs of insect origin used by aborigines of Rajasthan State, India. *Ann. Ent.,* 8 (2) 43-45.

Singh, R. P. 1978. Present status of antifeedant for protection of Agricultural crops and possibilities of development in future. *Proc. Mtg. Insecticides of Plant Origin,* New Delh, pp. 112-115.

Singh, R. P. 1983. Search for antifeedant in some botanicals. *Z. Angew. Ent.* 96: 316-319.

Singh, R. P., Deva Kumar, C. and Dhingra, S. 1988. Activity of neem *(Azadirachta indica)* seed kernel extract against *Lipaphis erysimi Kalt. Phytoparasitica,* 16 (3) :225-230.

Singh, R. P. and Kataria, P. 1985. Toxicity of plant extracts to mosquito larvae. *Indian J. Ent.,* 47 (4): 401-404.

Singh, R. P. and Pant, N. C. 1980a. Lycorine a resistance factor in the plants of family Amaryllidaceae against desert locust, *Schistocerca gregaria* F. *Experientia,* 36: 552-553.

Singh, R. P. and Pant, N. C. 1980b. Investigation on the antifeedant property of subfamily Amaryilidoidea (Amaryllidaceae) against desert locust, *Schistocerca gregaria* F. *Indian J. Ent.,* 42 (3) : 465-468.

Singh, R. P. and Singh, S. 1985. Evaluation of deoiled neem *(Azadirachta indica* A. Juss) seed Kernel against *Trogoderma granarium* Eversts *Curr. Sci.,* 54 (18) : 950-951.

Singh, R. P., Singh, Y. and Singh, S. P. 1985, Field evaluation of neem *(Azadirachta indica* A. Juss) seed kernel extract against pod borers of pigeon pea *Cajanus Cajan* (L) Millap. *Indian J. Ent.* 47 (1) : 111-112.

Singh, R. P. and Srivastava, B. G. 1983. Aleohol extract of neem *(Azadirachta indica* A. Juss) seed oil as oviposition deterrent of *Dacus cucurbitae* (Coq.) *Indian J. Ent.,* 45 (4) : 497-498.

Singh, S. P., Sharma, P. and Vats, L. K. 1987. Light-dependent toxicity of the extract of plant *Tagetes erecta* and alpha-terthienyl towards larvae of mosquito *Culex tritaeniorhychus. Toxicological and Environmental Chemistry.* 16 (1):1-8.

Singh, V. S. and Ramachandran, R. 1981. Abstract of the paper presented at the National Seminar on *Strategies of Pest Management.* Entomological Society of India, New Delhi. Dec. 21-23, 1981.

Singh, Vijai, S.; Saxena, J. D.; Bhatia, S. K.; Phadke, K. G. and Singh, J. 1973. Severe incidence of shootfly on late-sown wheat. *Entomologists, Newslet.,* 3 (10): 63.

Sinha, M. M. and Singh, B. 1988. Madhumakhi Palan Se Labh (in Hindi). *Adhunik Kisan,* 18 (6) : 15-20.

Slone, T. H. 1991. Insects in Chinese Medicine. *The Food Insects Newslet.,* 4 (3) :5.

Snodgrass, R. E. 1952. *A Text Book of Arthropod Anatomy.* Ithaca, New York, 366 pp.

Srivastava, P. D. 1953. Agricultural products as insecticides. *Sci. and Cult.,* 19: 139-140.

Srivastava, P.D. 1968. Role of Hermit crabs in the biological control of *Achatina fulica* Bowdich. on the Andamans. *Indian J. Ent.,* 30 (3) : 217-219.

Srivastava, P.D.and Srivastava, Y.N. 1967. *Orthomorpha* sp. a new Predatory Millipede on *Achatina fulica* in Andamans. *Experientia,* 23:776-777.

Stebbing, E. P. 1914. *Indian Forest Insects of Economic Importance—Coleoptera.* Eyre and Spottiswoode Ltd., London, 648 pp.

Storer, T. I. and Usinger, R. L. 1965. *General Zoology.* McGraw Hill Book Co., Inc. New York, London. 741 pp.

Sukhani, T.R. 1986. Role of insecticides in sorghum insect pest management. *Pestology,* 10 (4) : 17-23.

Teotia, T.P.S. and Tewari, G. C. 1971. Dharek drupes and leaves as protectant against *Sitotroga cerealella* infesting wheat seeds. *Bull. Grain Tech.,* 9 (1): 7-12.

Thomas, J. Muzik, 1970. *Weed Biology and Control.* McGraw Hill Book Co., New York. 273 pp.

Tiegs, O. W. 1940. The embryology and affinities of the symphyla based on a study of *Hansenielle agilis. Q. J. M. S.* 82: 1-225.

Tiegs, O. W. 1945. The postembryonic development of *Hanseniella agilis* (Symphyla). *Q.J.M.S.* 85:191-328.

Vishnu Swarup, 1980. Ornamentals. In: *Handbook of Agriculture :* (Ed). S. N. Tata. Indian Council of Agricultural Research, New Delhi, pp. 1049-1127.

Wadhi, S. R. 1983. Plant Quarantine in India. In: *Agricultural Entomology I.* (Eds) P. D. Srivastava, M.G. Jotwani, R. A. Agarwal, S.R. Wadhi, R.K. Bhanotar, R.K. Bhatnagar. All India Scientific Writers' Society, New Delhi, pp. 158-171.

Wilson, M.C., Broersma, D.B. and Provonsha, A. V. 1984. *Fundamentals of Applied Entomology.* Waveland Press, Inc. Prospect Heights, Illinois, U.S.A., 216 pp.

INDEX